Wine Grape Varieties

in California

LARRY J. BETTIGA
University of California Cooperative Extension Viticulture Farm Advisor;
Monterey, San Benito, Santa Cruz Counties

L. PETER CHRISTENSEN
Cooperative Extension Viticulture Specialist Emeritus, UC Davis Department
of Viticulture and Enology

NICK K. DOKOOZLIAN
University of California Cooperative Extension Viticulture Specialist,
UC Davis Department of Viticulture and Enology

DEBORAH A. GOLINO
University of California Cooperative Extension Plant Pathology Specialist
and Director, UC Davis Foundation Plant Services

GLENN MCGOURTY
University of California Cooperative Extension Viticulture Farm Advisor;
Lake, Mendocino Counties

RHONDA J. SMITH
University of California Cooperative Extension Viticulture Farm Advisor,
Sonoma County

PAUL S. VERDEGAAL
University of California Cooperative Extension Viticulture Farm Advisor,
San Joaquin County

M. ANDREW WALKER
Professor and Geneticist, Agricultural Experiment Station–UC Davis
Department of Viticulture and Enology

JAMES A. WOLPERT
University of California Cooperative Extension Viticulture Specialist,
UC Davis Department of Viticulture and Enology

EDWARD WEBER
University of California Cooperative Extension Viticulture Farm Advisor,
Napa County

UNIVERSITY OF CALIFORNIA
Agriculture and Natural Resources
Publication 3419

For information about ordering this publication, contact

University of California
Agriculture and Natural Resources
Communication Services
6701 San Pablo Avenue, 2nd Floor
Oakland, California 94608-1239

Telephone 1-800-994-8849
(510) 642-2431
FAX (510) 643-5470
E-mail: danrcs@ucdavis.edu
Visit the ANR Communication Services web site at
http://anrcatalog.ucdavis.edu

Publication 3419

UC PEER REVIEWED This publication has been anonymously peer reviewed for technical accuracy by University of California scientists and other qualified professionals. This review process was managed by the ANR Associate Editor for Pomology, Viticulture, and Subtropical Horticulture.

ISBN 1879906635

Library of Congress Control Number: 2003104162

 Printed in Canada on recycled paper.

Contents

Acknowledgments

This publication would not have been possible without the talented and technical contributions of authors, reviewers, and photography, design, and editorial staff. A special thank you is due to those viticulture farm advisors who served as authors and reviewers. Credit for the exceptional photography goes to Jack Kelly Clark, principal photographer at ANR Communication Services. Will Suckow, design project manager at ANR Communication Services, is responsible for the beautiful design. Ann Senuta, publications manager at ANR Communication Services, provided the motivation, coordination, and production expertise. Growers, vineyard managers, equipment manufacturers, and winemakers were widely consulted for information on variety performance and characteristics, cultural practices, machine harvestability, and wine-making considerations. Their invaluable inputs are greatly appreciated.

—*The Editors*

Wine Grape Varieties

in California

Introduction

California's first grape growers and winemakers were the Mission fathers in the 1770s. From those earliest plantings until nearly the middle of the nineteenth century, grape production at the California missions and in commerce was of a single variety, Mission. This changed rapidly after the Gold Rush and during the 1850s to 1880s when nurseries and growers began importing many European varieties from New England and the Old World. Wine industry pioneers responsible for variety introductions include Antoine Delmas (1852), Louis Pellier (early 1850s), Charles Lefranc (1858), Emil Dresel (1859), Agoston Haraszthy (1862), Hiram Crabb (late 1860s), Jean-Baptiste Portal (1872), Charles Wetmore (late 1870s), John Drummond (1880), John Doyle (1880s), and Charles McIver (1880s). The State Viticultural Commission, established in 1860 by an act of legislature, was instrumental in promoting wine variety importation as well as providing information to the emerging industry.

Research on the suitability of wine varieties to different regions began in California with the arrival of Professor Eugene Hilgard at the University of California, Berkeley, in 1874. Hilgard insisted that high wine quality should be the primary and long-term goal, including selection of the best varieties. About 1900, Frederic Bioletti assumed the major viticulture responsibilities and also began planting vineyards at the new University Farm at Davis (1908). He prepared one of the earliest California ampelographic works in the 1930s and conducted variety and rootstock trials throughout the state. Albert Winkler arrived in Davis in 1920, followed by Harry Jacob (1921), Harold Olmo (1934), Maynard Amerine (1935); all had profound influences on wine variety introductions, evaluation, and recommendations. Their work laid the foundation for recommendations of wine varieties best suited for the state's varying climatic regions.

Continued variety and clone introductions, the breeding of new varieties, and variety and winemaking trials from all regions in the state supported the effort.

The wine boom that began in the 1960s forever changed the mix and diversity of wine varieties in California. Consumer and marketing trends rapidly shifted away from dessert and generic wines to quality table wines, particularly varietal wines. Winemaking technology, knowledge, and experience improved wine quality and the range of styles. New viticultural areas were established, many of which are now major districts. Vineyard establishment and cultural practices improved greatly through technology and innovation. Important contributions during this period include the development and use of drip irrigation, improved trellis design and canopy management practices, implementation of integrated pest management, rootstock and clonal selection, virus detection and elimination, and mechanization.

The California wine industry continues to transform the vineyard landscape by developing new regions and experimenting with new varieties. Many growers and vintners throughout the state are planting and evaluating varieties that are lesser known to California, notably those from southern France, Spain, and Italy.

The purpose of *Wine Grape Varieties in California* is to serve as a teaching tool, a reference, and a guide to growers and vintners making decisions about variety plantings and growing practices. It is the first comprehensive variety publication to serve all wine growing districts in California. It should be a useful reference for other wine grape growing regions as well.

Wine Grapevine Structure

Typical vinifera grape leaf with five lobes

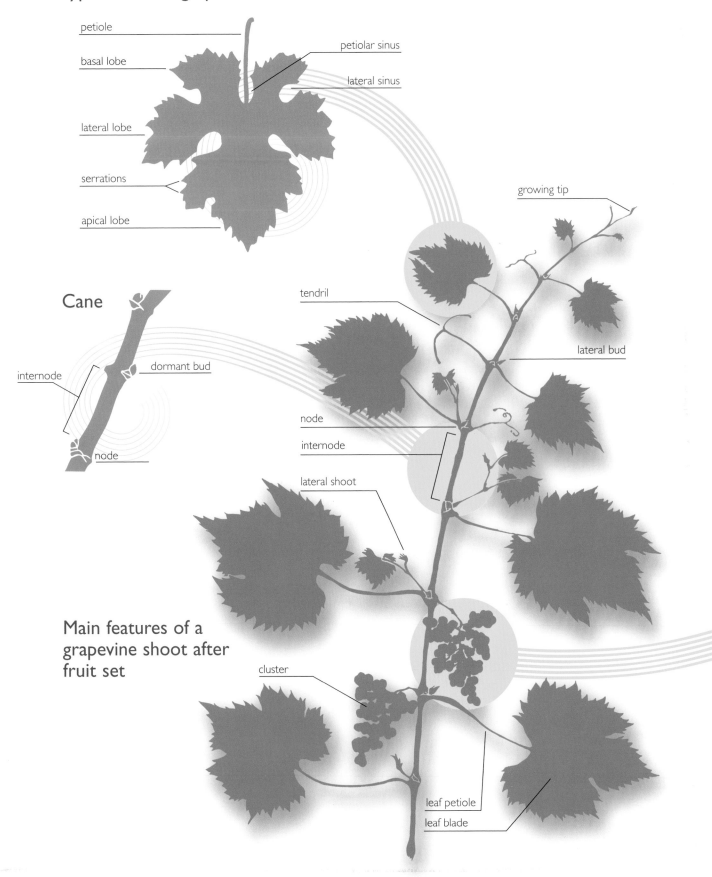

petiole

petiolar sinus

basal lobe

lateral sinus

lateral lobe

serrations

apical lobe

growing tip

Cane

tendril

internode

dormant bud

lateral bud

node

node

internode

lateral shoot

Main features of a grapevine shoot after fruit set

cluster

leaf petiole

leaf blade

Grape cluster and its attachment to cane

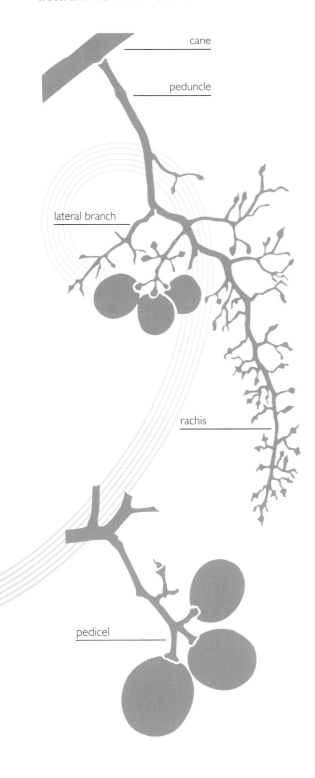

cane

peduncle

lateral branch

rachis

pedicel

Cluster and Berry Size and Shape

Common grape cluster shapes

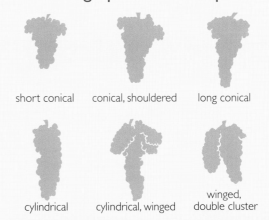

short conical   conical, shouldered   long conical

cylindrical   cylindrical, winged   winged, double cluster

Cluster Weight (lb)*	Class
<.25	small
.25 to .33	medium–small
.33 to .50	medium
.50 to .67	medium–large
.67 to .85	large
>.85	very large

*Average cluster weights are not described in each variety profile due to the variability among clones, rootstocks, soil conditions, districts, cultural practices, and seasonal weather.

Grape berry shapes

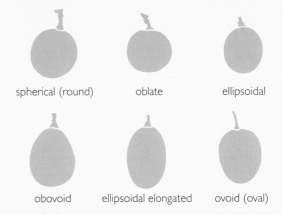

spherical (round)   oblate   ellipsoidal

obovoid   ellipsoidal elongated   ovoid (oval)

Berry Weight (g)	Class
<1.4	small
1.4 to 1.7	medium–small
1.7 to 2.0	medium
2.0 to 2.4	medium–large
2.4 to 3.0	large
>3.0	very large

Ripening Periods
of California Wine Grape Varieties

Early	Early-Mid	Mid	Mid–Late	Late
Chardonnay	Arneis*	Barbera	Alicante Bouschet*	Aglianico*
Gamay noir	Chenin blanc	Burger	Cabernet franc	Carignane
Melon	Gewürztraminer	Carnelian*	Cabernet Sauvignon	Montepulciano*
Muscat blanc	Malvasia bianca	Centurion*	Cinsaut*	Mourvèdre
Orange Muscat*	Roussane	Colombard	Dolcetto*	Muscat of Alexandria
Pinot noir	Semillon	Freisa*	Durif	Petit Verdot*
Sauvignon blanc	Sylvaner*	Grenache	Malbec	Mission*
Viognier	Syrah	Marsanne*	Nebbiolo	Rubired
	Tempranillo	Merlot	Tannat*	Ruby Cabernet
	Tinta Madeira*	Riesling	Valdiguié	
	Trousseau gris*	Sangiovese		
		Symphony*		
		Ugni blanc*		
		Zinfandel		

*Minor varieties

Ripening Dates

of California Wine Grape Varieties by Growing District

District/County	Early	Early–Mid	Mid	Mid–Late	Late
Anderson Valley/Mendocino	Sparkling: Sept. 1–15 Still: Sept. 15–Oct. 15	Sept. 15–Oct. 10	Oct. 1–20	Oct. 21–Nov. 21	NA
Carneros/Napa–Sonoma	Sparkling: Aug. 25–Sept. 15 Still: Sept. 5–25	Sept. 7–30	Sept. 15–Oct. 15	Sept. 25–Oct. 15	NA
Fresno/Fresno	Aug. 10–31	Aug. 20–Sept. 10	Sept. 5–30	Sept. 15–Oct. 5	Sept. 25–Oct. 25
Healdsburg/Sonoma	Sept. 1–15	Sept. 7–21	Sept. 15–30	Sept. 21–Oct. 31	NA
Lodi/San Joaquin	Aug. 15–31	Sept. 1–15	Sept. 10–25	Sept. 20–Oct. 5	Sept. 25–Oct. 25
Los Olivos/Santa Barbara	Sept. 1–15	Sept. 1–20	Sept. 7–25	Sept. 15–Oct. 7	NA
Paso Robles/San Luis Obispo	Sept. 1–15	Sept. 5–20	Sept. 10–25	Sept. 25–Oct. 15	Oct. 1–31
Plymouth/Amador	Sept. 1–15	Sept. 5–30	Sept. 20–Oct. 25	Oct. 1–25	Oct. 1–31
San Lucas/Monterey	Sept. 1–30	Sept. 5–Oct. 5	Sept. 25–Oct. 15	Oct. 1–31	NA
St. Helena/Napa	Sept. 1–15	Sept. 5–25	Sept. 15–Oct. 7	Sept. 20–Oct. 15	Oct. 1–31
Soledad/Monterey	Sparkling: Aug. 25–Sept. 15 Still: Sept. 15–Oct. 15	Sept. 25–Oct. 15	Oct. 5–31	Oct. 20–Nov. 7	NA
Ukiah/Mendocino	Sparkling: Aug. 20–31 Still: Sept. 1–20	Sept. 5–30	Sept. 21–Oct. 15	Oct. 15–Nov. 14	Oct. 21–Nov. 21

Vine Selection and Clones

During the 1930s and 1940s, it became clear that virus diseases were reducing the productivity and quality of some vineyards in California. In addition, many commercially available grapevine selections were mislabeled or incorrectly identified. By 1952, Harold Olmo led the formation of the California Grape Certification Association to develop, maintain, and distribute virus-tested and correctly identified grape stock. By 1958, this program combined with the UC Davis disease-tested fruit and nut tree program to become Foundation Plant Materials Service. Curtis J. Alley, a student of Dr. Olmo's who went on to become a faculty member in the UC Davis Viticulture and Enology Department, initially managed the collection. William Hewitt and Austin Goheen, UC Davis and USDA-ARS plant pathologists respectively, provided expertise in virus detection and elimination.

This program is now known as Foundation Plant Services (FPS). It is the home of the Foundation Vineyard, the source of grape varieties in the California Grapevine Registration and Certification (R & C) Program. The vines at FPS are "registered vines" with the California Department of Food and Agriculture. They must be maintained at certain standards of disease testing and inspection under state regulation. Wood from this Foundation Vineyard is sold to grapevine nurseries in the R & C program where it is also registered and maintained under a set of regulations. These blocks of grapevines can be used by nurseries to create "certified stock" sold to growers. This certification assures growers that vines have successfully completed extensive virus testing. The majority of California's grape planting stock originates from this program. FPS is also authorized to import new grape selections from around the world, which adds to the diversity of planting stock available to grape growers. In addition to importing clones, FPS also works to preserve clones growing in California's premier vineyards.

There are two critical areas that need to be considered in developing a superior grape variety collection. First is disease status. Until a new selection is free of virus, vine performance is impossible to evaluate because vigor, yield, and fruit quality are all affected by grapevine viruses. By using certified grape nursery stock, growers can reduce uncertainty about vine performance. Secondly, as selections of the same variety from different sources are compared, subtle performance differences between selections of the same wine grape variety become apparent. These differences are caused by mutations in genes that control characters such as leaf lobing, berry color, disease resistance, and ripening date. Over time, mutations accumulate and lead to greater diversity in older varieties. Selections that differ in these ways and have been evaluated are known as "clones" of a variety. Planting superior clones can improve a variety's production and winemaking characteristics.

Today, with increasingly diverse plant materials available, growers planting new vineyards need to consider choice of clone as well choice of variety. Most of the older FPS selections were collected by UC Davis scientists over the years both by selection from superior California vineyards and by plant exploration in other countries. New clones continue to originate from formal clonal selection programs and public research projects around the world. Some of the programs that have contributed significantly to clonal diversity in California are ENTAV (Etablissement National Technique pour l'Amélioration de la Viticulture, in France); Geisenheim (Geisenheim Research Institute, in Germany); and Rauscedo (Vivai Cooperativi Rauscedo, in Italy). Where appropriate, the clone numbers of these programs as well as the selection numbers used by FPS are provided in the individual variety sections of *Wine Grape Varieties in California.*

New clones of the major wine grape varieties are added to the FPS Foundation Vineyard frequently. Researchers, viticulturists, and winemakers around the state work to ensure that valuable "heritage" field selections—those collected from premier vineyards with a reputation for quality wine—are available as certified selections. In some of California's oldest vineyards, these selections represent pre-1900 European introductions that may contribute greatly to varietal clonal diversity.

The same clone can be introduced more than once to FPS; each introduction receives a unique selection number to preserve its identity. In addition, sub-clones that have been produced by heat treatment or tissue-culture virus-elimination therapy also receive unique numbers. This has led to a sometimes bewildering accumulation of selection and clone numbers for plant materials that may not differ significantly in performance. This is even further complicated by the existence of European clones that have reached California through other importation centers and may be named according to a number of different conventions.

An additional complication results from the intellectual property issues that have developed around wine grape clones. Some clones are

trademarked and/or proprietary while others are in the public domain. For example, FPS has public or "generic" selections of many of the ENTAV clones in the Foundation Vineyard collection that are known as "Reported to be French" with an ENTAV-assigned clone number as well as an FPS selection number. More then one FPS selection may be available from the same French clone, due to independent import and subcloning. These selections have no assurance of authenticity. However, clones that have come directly from ENTAV are part of a trademark program and are known as "ENTAV-INRA®." Not all of these trademarked ENTAV-INRA® clones are California certified; some have origins independent of the FPS program.

Today in California there is an unprecedented wealth of clonal material of the major grape varieties. Decisions on clones have become an integral part of the vineyard planning process. As in other wine regions, California growers want to know how clones might enhance viticultural performance and wine quality or help create a particular wine style. Along with this heightened interest in clones, several important points must be kept in mind:

Clone choice is only one of many important decisions when establishing a vineyard.
Variety choice, site climate, soil type, vineyard design (spacing, trellising, and rootstock), and annual cultural practices (irrigation, canopy management, and crop load) will impact final wine quality far more profoundly than clonal choice. There is no such thing as a "perfect" clone that will overcome a grower's inappropriate site selection or poor management decisions.

There is no one "best" clone. A clone's suitability for a particular vineyard depends on the target wine market and desired wine style, as well as the site and vineyard conditions noted above. High-yielding clones are just as appropriate for low-cost wines as low-yielding clones are for high-value wines. When the retail bottle price for a variety can vary by more than 20-fold, there is clearly room for more than one clone. Thus, the term "best" is value laden and must be carefully defined by the producer's goals.

A clone selected in another country is not necessarily superior to what is available locally. Clones from regions known for fine wine, particularly Burgundy and Bordeaux, are highly sought after. However, the selection criteria must be explicitly understood to ascertain whether clones selected abroad have value in California. Climates abroad can differ dramatically from California's. Some wine regions have strict yield limits, often made possible by the high bottle prices their wines command. Thus, "good" clones in those regions are those that perform well under specific environmental and economic conditions, neither of which may apply to California. Before making a significant investment in clones selected in another country, growers should verify the performance of those clones under their local conditions.

Virus infections can compromise even the best clones. Clones selected for high-quality wine and freedom from virus are the most desirable. Severe viruses are not tolerated in any of the world's clonal selection programs. Growers should be aware that, in addition to registered selections of clones, there is a great deal of common stock sold in California that has never been checked for virus. European clonal selections that have entered the United States illegally and field selections from old, established vineyards are frequently infected by virus. Growers should avoid virus-infected planting stock since many commonly used rootstocks are very sensitive to virus diseases.

Continued clonal evaluation of major varieties such as Cabernet Sauvignon, Chardonnay, and Zinfandel in California is supported by growers' and vintners' funding organizations. However, this progress is challenged by several complicating factors:

First, many varieties contribute significantly to the state's wine economy. In other countries, regions can concentrate on relatively few varieties of importance. Even for "niche" varieties with small acreage such as Malbec, Viognier, Sangiovese, or Tempranillo, clonal performance is a significant issue because a new variety will not be accepted if the principal clone in use performs poorly.

Secondly, fine wine production in California spans a wide variety of climatic regions, elevations, and soil types. For the major varieties—Chardonnay, Cabernet Sauvignon, Merlot, Zinfandel, Pinot noir—multiple trials are needed to understand how clones will perform in different regions. In addition, the flood of new clonal material in California brings a continuing need for new sets of trials.

Finally, clonal evaluation involves three consecutive steps: viticultural analysis (growth and yield components, rot susceptibility), wine analysis (chemical data, color, tannin), and sensory evaluation. Each step builds on the previous one, but with increasing resource needs. Evaluating the great diversity of clonal material now available in California will require expanding current viticultural research programs and winemaking evaluations.

—Deborah A. Golino and James A. Wolpert

Rootstock Selection

Almost all the wine varieties described in *Wine Grape Varieties in California* are of pure *Vitis vinifera* parentage. This species is particularly prone to attack by two root pests: grape phylloxera and parasitic nematodes. *Vitis vinifera* vines can be protected from these pests by grafting them to rootstock varieties derived from other vine species and resistant hybrids. Many of the rootstocks used for this purpose are adapted to particular soil types, chemistry, and fertility. They may also be used to overcome vineyard problems such as drought, excess water, and salinity.

It is important that growers select rootstocks that are:
- resistant to present and potential soil pests
- suitable for the soil's texture, depth, and fertility
- compatible with soil chemistry (pH, salinity, lime content)
- favored for the anticipated soil water availability, drainage, and irrigation practice
- appropriate for the vineyard design, and
- appropriate for the fruiting variety's growth and fruiting characteristics.

Rootstock	Vitis parentage	Phylloxera resistance	Nematode Resistance		Tolerance			
			Root knot	Dagger (*Xiphinema index*)	Drought	Wet soil	Salinity	Lime
Riparia Gloire	*riparia*	High	Low	Med.	Low	Low	Med.	Low
St. George (*Rupestris du lot*)	*rupestris*	High	Low	Low	Low–med. in shallow soils; high in deep soils	Low–med.	Med.–high	Med.
SO4 (Selection Oppenheim)	*berlandieri* × *riparia*	High	Med.–high	Low–med.	Low–med.	Med.–high	Low–med.	Med.
5BB (Kober)	*berlandieri* × *riparia*	High	Med.–high	Med.	Med.	Low	Med.	Med.–high
5C (Teleki)	*berlandieri* × *riparia*	High	Med.–high	Low–med.	Low	Low–med.	Med.	Med.
420A (Millardet et de Grasset)	*berlandieri* × *riparia*	High	Med.	Low	Med.	Low–med.	Low	Med.–high
99R (Richter)	*berlandieri* × *rupestris*	High	Med.–high	Low–med.	Med.–high	Low	Med.	Med.
110R (Richter)	*berlandieri* × *rupestris*	High	Low–med.	Low	High	Low–med.	Med.	Med.

In this publication, each wine variety's description includes information on rootstock suitability and experience. Additionally, the following table provides comparative information that may assist growers with rootstock decisions. This information is based on written reports throughout Europe and the New World, including California. The ratings and comments provided here are based on prevalence in the literature and those reports most likely to fit California conditions. As more experience is gained in California's diverse viticulture environment, deviation from the information in the table can be expected. Widespread University of California and grower rootstock trials will continue to provide updated information on rootstock performance. Growers should also consult with their Cooperative Extension farm advisor, other local growers, consultants, and nursery representatives when selecting a rootstock.

—L. Peter Christensen

Influence on scion		Soil adaptation	Ease of propagation	Other characteristics
Vigor	Mineral nutrition[1]			
Low–med.	N, P: low K, Mg: low–med.	Deep, well-drained, fertile, moist soils	High	Early maturation; scions tend to overbear
High	N: high P: low on low-P soils, high on high-P soils K: high	Deep soils	High	Fruit set problems with some scions; latent virus tolerant
Low–med.	N: low–med. P: med. K: med.–high Mg: med.	Moist, clay soils	Med.	Noted as a cool-region rootstock
Med.	N: med.–high P, K, Zn: med. Ca, Mg: med.–high	Moist, clay soils	High	Susceptible to phytophthora root rot; adapted to high-vigor varieties
Low–med.	N: low P, K: med. Mg: med.–high Zn: low–med.	Moist, clay soils	High	—
Low	N, P, K: low Mg: med. Zn: low–med.	Fine-textured, fertile soils	Med.	Scions tend to overbear when young
Med.–high	P: med. K: high Mg: med.	Tolerant of acid soil	Med.	Young scions may develop slowly
Med.	N: med. P: high K: low–med. Mg, Zn: med.	Hillside soils; acid soils	Low–med.	Develops slowly in wet soils

[1]Influence on scion mineral nutrition refers to comparative petiole tissue levels of nutritional elements.
— not available

Rootstock Selection *(continued)*

Rootstock	Vitis parentage	Phylloxera resistance	Nematode Resistance		Tolerance			
			Root knot	Dagger (*Xiphinema index*)	Drought	Wet soil	Salinity	Lime
140Ru (Ruggeri)	*berlandieri* × *rupestris*	High	Low–med.	Low	High	Low	Med.–high	Med.–high
1103P (Paulsen)	*berlandieri* × *rupestris*	High	Med.–high	Low	Med.–high	Med.–high	Med.	Med.
3309C (Couderc)	*riparia* × *rupestris*	High	Low	Low	Low–med.	Low–med.	Low–med.	Low–med.
101-14 Mgt (Millardet et de Grasset)	*riparia* × *rupestris*	High	Med.–high	Med.	Low–med.	Med.	Med.	Low–med.
Schwarzmann	*riparia* × *rupestris*	High	Med.	High	Med.	Med.	Med.–high	Med.
44-53M (Malègue)	*riparia* × (*cordifolia* × *rupestris*)	High	Low	—	High	—	—	Low–med.
1616C (Couderc)	*longii* × *riparia*	High	High	Med.	Low	High	Med.–high	Low–med.
Salt Creek (Ramsey)	*champinii*	High	High	Low–med.	Med.–high	Low–med.	High	Med.
Dogridge	*champinii*	Med.	Med.–high	Low–med.	Med.	Low–med.	Med.–high	Med.
Harmony	1613 (*solonis* × *Othello*) × Dogridge	Low–med.[2]	Med.–high	Med.–high	Low–med.	Low	Low–med.	Med.
Freedom	1613 (*solonis* × *Othello*) × Dogridge	Low–med.[2]	High	High	Med.	Low	Low–med.	Med.
O39-16	*vinifera* × *rotundifolia*	High	Low	High	Low	—	Low	Low

Influence on scion		Soil adaptation	Ease of propagation	Other characteristics
Vigor	Mineral nutrition[1]			
High	N: med.–high P, Mg: high K: low	Adapted to drought and acid soils	Med.	Does poorly in non-irrigated, low K soils
Med.–high	N: med.–high P, Mg: high K, Zn: low–med.	Adapted to drought and saline soils	High	—
Low–med.	N: med.–high P, Ca: low K, Mg, Zn: med.	Deep soils	High	Sensitive to latent viruses; tolerant of cold injury
Med.	N, K: med.–high P, Mg, Ca: low Zn: med.	Moist, clay soils	High	—
Med.	N, P: med. K: med.–high Mg: low	Moist, deep soils	High	—
Med.	N: low–med. P, Mg, Ca: low K: high	High Mg soils	High	Readily Mg deficient in low Mg soils
Low	N: low K: med.–high	Best on fertile, med.- to fine-textured soils	High	Poor on low-vigor sites; tolerates wet soils
High	N, P: high K: med.–high Zn, Mn: low	Sandy, infertile	Low	Tolerant to *Phytophthora*
Very high	N, P: high K: med. Zn: low	Very sandy, infertile	Low	Promotes excess vigor, poor fruit set
Med.–high	N: low P: med. K: high Zn: low–med.	Sandy loams and loamy sands	High	—
High	N, P, K: high Mg: med. Zn, Mn: low	Sandy to sandy loams	Med.–high	Sensitive to latent viruses
High	N, K: high P: low–med. Zn: low	Poor on coarse, sandy soils due to low root knot nematode tolerance	Very low	Tolerant of fanleaf virus

[1] Influence on scion mineral nutrition refers to comparative petiole tissue levels of nutritional elements.
[2] The degree of long-term phylloxera resistance is questionable due to the unknown *Vitis vinifera* parentage of these rootstocks.
— not available

Trellis Selection and Canopy Management

Over the past two decades, advancements in vineyard design, trellis and training systems, and canopy management practices have dramatically improved wine grape productivity and fruit quality in California. Prior to this period, a standard vineyard design and trellis system was used throughout the state. Little attention was paid to site-specific factors influencing vine vigor such as climate, growing region, soil type, and rootstock. Now significant effort is made to match vineyard design and trellis system to the site-specific factors that influence potential vine growth. As a result, a wide range of plant densities and training/trellis systems are routinely employed in California wine grape production. The trellises used range from single to divided curtain systems and employ both horizontal and vertical canopy division. Due to both cost and durability, metal has replaced wood as the material of preference for trellis construction.

The major wine grape trellis systems currently used in California are outlined in the following table. A primary consideration when selecting the proper trellis system is anticipated vine vigor or canopy size. Highly

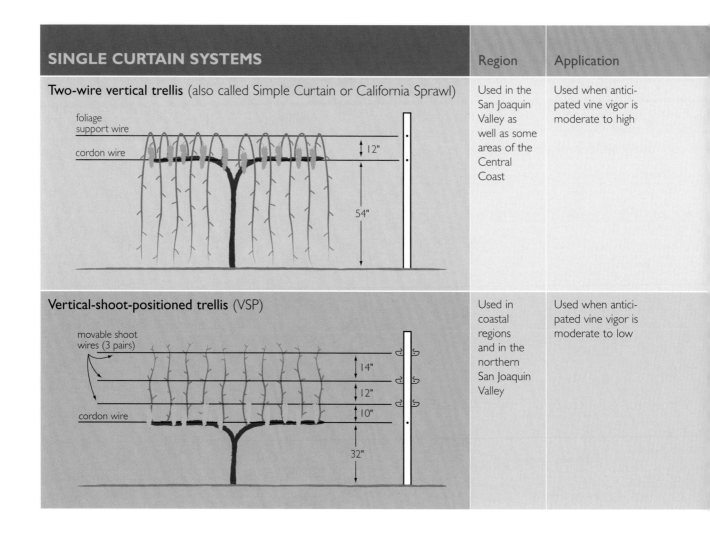

SINGLE CURTAIN SYSTEMS	Region	Application
Two-wire vertical trellis (also called Simple Curtain or California Sprawl) foliage support wire cordon wire 12" 54"	Used in the San Joaquin Valley as well as some areas of the Central Coast	Used when anticipated vine vigor is moderate to high
Vertical-shoot-positioned trellis (VSP) movable shoot wires (3 pairs) 14" 12" cordon wire 10" 32"	Used in coastal regions and in the northern San Joaquin Valley	Used when anticipated vine vigor is moderate to low

vigorous vines require larger, more expansive trellising systems than low-vigor vines. Before vineyard establishment it is important to accurately estimate anticipated vine vigor or canopy size to select the proper trellis system.

Climate plays a major role in determining vine growth potential, particularly temperature, annual rainfall, sunlight exposure, and wind velocity. Warm summer temperatures and large amounts of sunlight exposure encourage large canopies, while cooler temperatures or constant and high-velocity winds in the spring and summer result in less-vigorous growth. Soil texture and potential vine-rooting depth also influence vine growth. Deep, fertile soils with large amounts of stored soil moisture support vigorous vine growth, while soils of moderate rooting depth and lower amounts of stored water support less growth. Lastly, pre-plant soil preparation (ripping or slip plowing), cultivar, rootstock selection, and anticipated cultural practices (irrigation, fertilization, and vineyard floor management) also impact vine growth.

Other factors influencing trellis choice include plant and row spacing, row orientation, establishment costs, equipment requirements, and the desire to mechanize labor-intensive practices such as pruning and harvesting.

Training and pruning systems	Spacing	Mechanization	Approximate cost	Comments
Bilateral cordon training and spur pruning	Spacing between vines is generally 6 to 8 feet. Spacing between rows is 10 to 12 feet.	Harvest; pruning or pre-pruning; leaf removal; hedging	$1,500 per acre for materials, trellis, and installation	Most common system for wine grape production in San Joaquin Valley due to low establishment cost and ease of pruning and harvest mechanization. Canopy configuration prevents excessive fruit sunlight exposure in warm climates. Interior canopy shading can be a problem if vines are highly vigorous.
Bilateral cordon training and spur pruning most common. Unilateral cordon training and spur pruning used when in-row vine spacing is 5 feet or less. Head training and cane pruning used for some cultivars in cool regions.	Spacing between vines is 3 to 8 feet. Spacing between rows is between 7 and 8 feet.	Harvest; pre-pruning; shoot positioning; leaf removal; hedging	$2,500 per acre for materials, trellis, and installation	Most common trellis system for wine grape production in coastal regions. Allows reduced between-row spacing and increasing vineyard design efficiency; requires shoot positioning.

In addition to proper trellis selection, canopy management practices such as basal leaf removal, shoot positioning, and hedging are an integral part of high-quality wine grape production. Some form of basal leaf removal is practiced in the majority of coastal wine grape vineyards, as well as in many vineyards in the northern and central San Joaquin Valley. Shoot positioning is per-

formed in all vineyards trellised to the lyre, vertical-shoot-positioned, Scott Henry, Smart-Henry, and Smart-Dyson systems. Some form of hedging or shoot trimming is necessary with most of these systems as well.

Basal leaf removal consists of removing primary leaves and lateral shoots that subtend the four to six basal nodes on each primary shoot. In

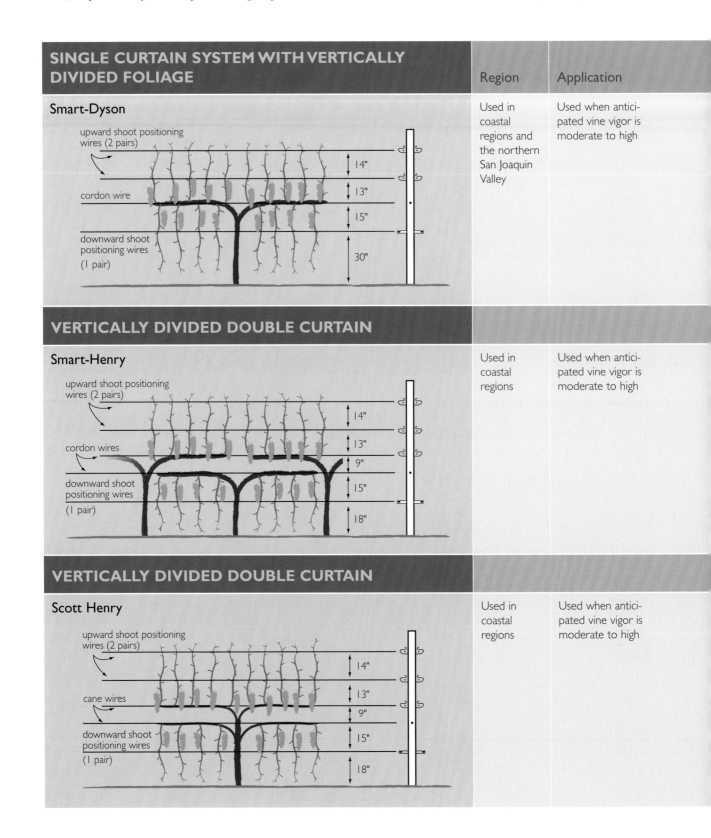

SINGLE CURTAIN SYSTEM WITH VERTICALLY DIVIDED FOLIAGE	Region	Application
Smart-Dyson	Used in coastal regions and the northern San Joaquin Valley	Used when anticipated vine vigor is moderate to high
VERTICALLY DIVIDED DOUBLE CURTAIN		
Smart-Henry	Used in coastal regions	Used when anticipated vine vigor is moderate to high
VERTICALLY DIVIDED DOUBLE CURTAIN		
Scott Henry	Used in coastal regions	Used when anticipated vine vigor is moderate to high

most regions leaves are removed on the shaded side of the row only (that is, the north side of east-west-oriented rows or the east side of north-south-oriented rows). Normally leaves are removed shortly after berry set to allow clusters to acclimate to increased sunlight exposure and higher temperatures and to reduce the likelihood of sunburn. Growers should avoid removing

leaves immediately before berry softening, or veraison, as fruit grown in the canopy shade is highly susceptible to sunburn if suddenly exposed at this time.

In many coastal vineyards, shoots are thinned in the early spring to reduce shoot congestion and crop load. Sterile shoots, and in some cases cluster-bearing shoots from non-count nodes, are

Training and pruning systems	Spacing	Mechanization	Approximate cost	Comments
Bilateral cordon training and spur pruning	Spacing between vines is generally 6 to 8 feet. Spacing between rows is 7 to 8 feet.	Harvest; pre-pruning; leaf removal; hedging	$2,500 per acre for materials, trellis, and installation.	Used for new vineyards or as a retrofit for existing vineyards trellised to VSP. Generally used in coastal regions when anticipated vine vigor is too high for VSP but narrow-row spacing is desirable. Between-row spacing should not be less than 7 feet in order to prevent shading of lower portion of canopy. Requires both upward and downward shoot positioning. Popularity increasing.
Bilateral cordon training and spur pruning	Spacing between vines is 6 to 8 feet. Spacing between rows is 7 to 8 feet.	Harvest; pre-pruning; leaf removal; hedging	$2,500 per acre for materials, trellis, and installation.	Used in coastal regions when anticipated vine vigor is too high for VSP but narrow row spacing is desirable. Requires that bilateral cordon–trained vines be alternated at two heights to create upper and lower fruiting zones. Lower fruiting zone often becomes weak over time.
Head training and cane pruning	Spacing between vines is 6 to 8 feet. Spacing between rows is 7 to 8 feet.	Harvest; leaf removal; hedging	$2,500 per acre for materials, trellis, and installation.	Similar application as Smart-Dyson, except that cane pruning allows easier separation of canopy. Used when cane pruning, VSP canopy configuration, and narrow-row spacing are desired under moderate- to high-vigor conditions.

removed when the average shoot length is 6 to 8 inches. Shoot thinning increases light reaching the basal buds in the canopy interior. However, under moderate- to high-vigor conditions the effects of this practice on canopy microclimate may be temporary due to compensating lateral shoot growth.

In vertical-shoot-positioned (VSP) canopies, shoot positioning maintains canopy form and foliage separation in narrow-row spacings. On horizontally divided canopies (GDC or Wye), shoot postioning maintains canopy separation. It improves light penetration to the canopy interior, particularly in vigorous, horizontally divided vineyards where the row middle or area between the fruiting zones becomes shaded following fruit set. The vine foliage is separated or positioned using movable wires. On vertically divided systems, shoot positioning is performed several times per year, typically near bloom and following berry set. For horizontally divided systems, shoot positioning is normally performed once per year near bloom.

Hedging or shoot trimming maintains canopy shape, prevents shading, and facilitates cultivation and mechanization. The shoots of VSP canopies are trimmed when the foliage begins to grow over the positioning wires at the top of the

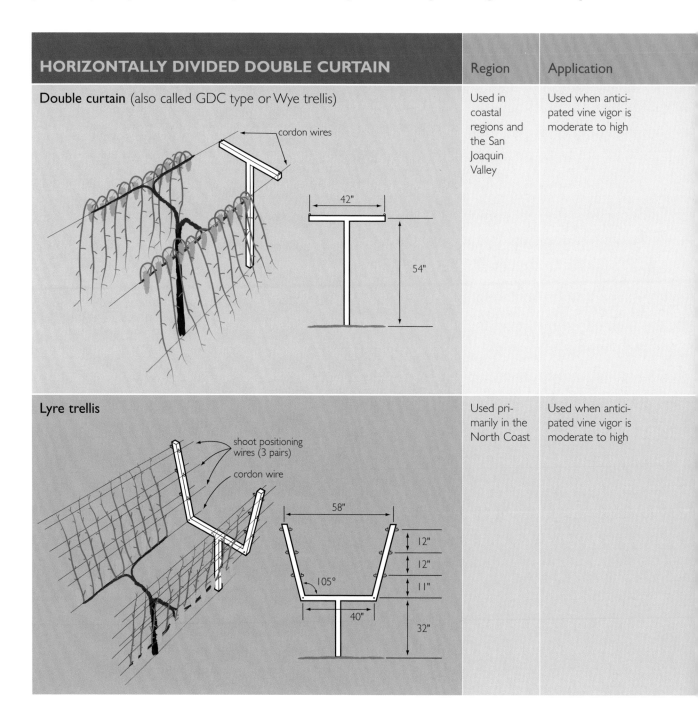

HORIZONTALLY DIVIDED DOUBLE CURTAIN	Region	Application
Double curtain (also called GDC type or Wye trellis)	Used in coastal regions and the San Joaquin Valley	Used when anticipated vine vigor is moderate to high
Lyre trellis	Used primarily in the North Coast	Used when anticipated vine vigor is moderate to high

canopy, usually sometime between berry set and veraison. The shoots are typically trimmed 6 to 8 inches above the top canopy wires. If significant lateral shoot growth has occurred, sides of the canopy are also hedged to maintain canopy width of approximately 18 to 20 inches. California Sprawl (two-wire vertical) canopies in the San Joaquin Valley are typically trimmed approximately 24 inches above the vineyard floor sometime near veraison. This facilitates air movement and decreases humidity in the fruiting zone.

—*Nick K. Dokoozlian*

Training and pruning systems	Spacing	Mechanization	Approximate cost	Comments
Quadrilateral cordon training and spur pruning; divided curtain may also be formed using bilateral cordon trained vines that alternate from side to side (e.g., GDC). Distance between curtains ranges from 2 to 4 feet, depending on desire to mechanize harvest.	Spacing between vines is 6 to 8 feet. Spacing between rows is 11 to 12 feet.	May be mechanically harvested if curtains are no more than 30" apart. Pruning or pre-pruning, hedging may also be mechanized.	$2,000 per acre for materials, trellis, and installation.	Used to reduce canopy density under high-vigor conditions. Shoot positioning used in some cases to increase sunlight penetration to the center of the canopy. Overcropping may be a problem with highly fruitful or large clustered cultivars.
Quadrilateral training and spur pruning; distance between curtains is 3 to 4 feet.	Spacing between vines is 6 to 8 feet. Spacing between rows is 10 to 12 feet.	Pre-pruning; leaf removal; hedging	$3,500 per acre for materials, trellis, and installation.	Found primarily in the North Coast in moderate- to high-vigor Cabernet Sauvignon and Merlot vineyards. Not widely used due to high establishment and annual production costs.

Major
Wine Grape
Varieties

in California

Barbera

Synonyms
Italian geographical names are sometimes used, such as d'Asti and del Monferrato.

Source
Barbera is a leading wine grape of Italy (second in planted acreage), particularly in the Piedmont region where it is thought to have originated. It is also important in Argentina and can be found in other South American countries as well as in Croatia. John Doyle first imported the grape into California and produced his first Barbera vintage in 1884 from vines planted in Cupertino. In the 1890s, the Italian Swiss Colony Winery used it successfully for several of its table wines. Yet it did not regain popularity after Prohibition until the rapid acreage expansion in the 1970s and 1980s, when it became a prominent red wine variety in the San Joaquin Valley, mostly for blending. In the coastal and foothill districts there is renewed interest in Barbera as a quality varietal wine grape and as a blend.

Description
Clusters: medium; conical, well filled to compact, can be winged; long peduncles.
Berries: medium; long oval, dark purple-black; relatively high acidity at maturity.
Leaves: medium; deeply 5-lobed, U-shaped petiolar sinus and superior lateral sinus that often overlap; relatively large, sharp teeth; wooly hair on lower surface.

Shoot tips: woolly and white with rose margin; youngest leaves have bronze-red highlights over lime-green background. Internodes are relatively long with straggly growth.

Growth and Soil Adaptability
The vine is moderately vigorous when grown on its own roots on medium- to fine-textured soils (sandy loam to clay loam); it is not vigorous enough on its own roots in sandier soils (loamy sands and sands). Its growth is trailing, and the vine canopy is somewhat open; its foliage is not dense except with extreme vigor. The canes are slender and attach themselves with strong tendrils, making pruning and brush removal from trellis wires difficult. The vine often produces a moderate second crop. Recommended in-row spacing is 6 or 7 feet. Single canopy vineyards in the San Joaquin Valley should have row spacing of 8 to 10 feet.

Rootstocks
Barbera has no known incompatibilities. Freedom and some Harmony rootstocks have been used in the San Joaquin Valley for nematode resistance and increased vine vigor in sandy loam and loamy sand soils. Ramsey may be needed in coarse, sandy soils. Phylloxera rootstocks successfully used in most California districts include Teleki 5C, Kober 5BB, 110R, 3309C, 101-14 Mgt, and 1103P; experience with other rootstocks was limited to older vineyards planted on their own roots or AXR #1 and St. George rootstocks. In Italy, 5BB and 420A are popular rootstocks for this variety.

clusters

Medium; conical, well-filled to compact, can be winged; long peduncles.

berries

Medium; long oval, dark purple-black; relatively high acidity at maturity.

Clones

Most of the plantings in the 1970s and 1980s were of Barbera FPS 01, also known as the Marshall (California) clone. Later this selection became non-registered due to the discovery of a leafroll virus in some wood sources. Subsequently, Barbera FPS 02, a virus-negative selection from Italy (Barbera Rauscedo 6) was registered. Barbera FPS 02 proved to be more fruitful and productive than selection 01, but it also produced larger berries and clusters that contributed to greater cluster compactness. A new virus-negative subclone of selection 01 that is smaller-berried and lower yielding is now available as Barbera FPS 06. It has fruit composition preferred to that of FPS 02. Recent clonal importation from Italian collections, including Barbera FPS 03 (Barbera CVT 171), FPS 04 (Barbera CVT 84), FPS 05 (Barbera CVT 171), FPS 07 (Barbera VCR 19), and FPS 08 (Barbera VCR 15) has increased diversity of Barbera planting selections in California.

Production

Vines usually bear 6 to 9 tons per acre, except in hillside and non-irrigated sites where lower yields are normal. Yields are also lower in the Sierra foothills, even in irrigated vineyards.

Harvest

Period: A midseason variety, harvested in mid-September to early October.

Method: The long, green peduncles make hand harvesting easy. Barbera is also one of the best varieties for machine harvesting. With canopy shaking fruit is easy to moderately easy to remove as single berries with some cluster parts. Juicing is medium. A trunk shaker removes fruit easily to moderately easily as single berries with some cluster parts and a few whole clusters. Juicing is light to medium.

Training and Pruning

Vines are most commonly trained to bilateral cordons and pruned to 12 to 16 spurs with two to three nodes each. Quadrilateral cordon training for crop load balance may be practiced where the vines are highly vigorous. Machine-hedge or non-selective pruning has been successful, resulting in increased yields with favorable fruit composition but some delay in fruit ripening.

leaves

Medium; deeply 5-lobed, U-shaped petiolar sinus and superior lateral sinus that often overlap; relatively large, sharp teeth; wooly hair on lower surface.

shoot tips

Woolly and white with rose margin; youngest leaves have bronze-red highlights over lime-green background. Internodes are relatively long with straggly growth.

Trellising and Canopy Management

The small to medium leaves and long internodes create a somewhat open canopy, exposing clusters and minimizing the need for canopy manipulation. San Joaquin Valley vineyards are most commonly trellised as a single curtain with the cordon wire at 42 to 54 inches. A single foliar support wire at 52 to 64 inches may be beneficial. Additional foliar wires are not recommended in warm districts or high-vigor sites. They increase the difficulty and cost of pruning due to the numerous strong tendrils that attach to foliar wires. GDC systems may benefit from foliage catch wires in warm districts. Vertical-shoot-positioned systems have also worked well even in the northern San Joaquin and Sacramento valleys. Lyre systems are sometimes used in vigorous coastal sites, but more acreage is head-trained in the Sierra foothills.

Insect and Disease Problems

Leafroll virus is moderate to severe in plantings before the early 1970s. The availability of heat-treated, virus-free selection FPS 01 in the early 1970s ended the problem in subsequent new plantings. Barbera is highly susceptible to Pierce's disease and very susceptible to downy mildew. Aerial crown gall following low temperature injury in the spring is common.

Other Cultural Characteristics

Barbera has low tolerance to sodic (alkali) or saline soils. Excessive flower shatter at bloom has occurred on vigorous vines with high nitrogen. Own-rooted vines tend to sucker at their base, and it is often necessary to remove watersprouts from trunks and cordons.

Winery Use

Barbera is popular in warm districts such as the San Joaquin Valley because of its high fruit acidity retention; it is mostly used for blending in such districts. In cooler regions and at lower yields it produces quality varietal wines of varying styles.

—*L. Peter Christensen*

Burger

Synonyms

In France, Burger is known as Monbadon, Grand blanc, Meslier d'Orleans, Castillone à Montendre, Caoba, and Elbling. Australian sources report Allemand, Bourgeois, and Mouillet as synonyms. In Germany, it has numerous synonyms, including Weiser, Elbling, Kleinberger, Grobriesling, Alben, Albig, Süssgrober, and Rheinelbe. In Portugal, it is called Alva; in Austria, it is named Grossriesling or Kurtzstingel; and in the former Yugoslavia the names Pezhech, Blesez, Morawkva, and Seretonia are used.

Source

The variety's source is uncertain, but there is some suggestion of introduction by Romans to Gaul or a local selection well-established before the Middle Ages. Its introduction to California is not documented.

Description

Clusters: large; long conical to cylindrical, compact; tip of cluster tends to be loose because of small berries; medium peduncles.

Berries: medium; round, yellow-white; very little pulp with high juice content.

Leaves: small; deeply 5-lobed, shoulder-like inferior lobes; petiolar sinus is narrow to closed U-shape; deep U-shape lateral sinuses; teeth somewhat rounded; rough, bumpy surface; tufted to dense hair on lower surfaces.

Shoot tips: felty white with red margin, young leaves downy with bronze highlights.

Growth and Soil Adaptability

Vines have moderate vigor and recumbent shoots. The variety grows well on fertile soils of medium texture; avoid extremely coarse soils if close management of water and nutrients is not possible. Vine spacing within rows is about 7 to 8 feet. Sandy soils with low zinc availability may increase flower set problems, and high productivity can induce potassium deficiency that resembles leafroll virus symptoms. Burger does not ripen well in cool climates.

Rootstocks

Moderate- to high-vigor rootstocks, such as Freedom and Ramsey, are usually suitable, especially on sandy, coarse soils where nematodes are a problem. The rootstocks 140Ru and 1103P are also possible.

Clones

Burger FPS 01, 02, 03, and 04 are registered selections; each originated from a different California vineyard and may be unique. Although little information on field performance or wine quality from fruit produced from these FPS selections is available, growers should be aware that many field selections have leafroll virus infections.

clusters

Large; long conical to cylindrical, compact; tip of cluster tends to be loose because of small berries; medium peduncles.

berries

Medium; round, yellow-white; very little pulp with high juice content.

Production

The vines are very productive relative to vine vigor, capable of consistently bearing large crops of 15 to 20 tons per acre.

Harvest

Period: A midseason variety, harvested in mid-September to early October.

Method: Hand harvest is easy due to large clusters and high productivity, with a tendency of berries to juice very easily. Trunk shaking is possible, but extra care is needed. Using a pivotal striker is not recommended due to the thin-skinned and juicy berries that can cause very high juice loss. The newer bow-rod heads may reduce this tendency.

Training and Pruning

Burger is well suited to spur pruning. If vertical cordons or bilateral cordons are used, 12 to 16 two-node spurs are acceptable, depending on rootstock, soil depth, and soil texture. Cane pruning may result in severe overcropping and reduced shoot vigor.

Trellising and Canopy Management

Bilateral or vertical cordon and spur pruning systems are used. Cane pruning is not recommended due to the variety's tendency to overcrop. Basal leaf removal may be utilized to reduce bunch rot at harvest. Shoot thinning may be employed on young vines to prevent overcropping.

leaves

Small; deeply 5-lobed, shoulder-like inferior lobes; petiolar sinus is narrow to closed U-shape; deep U-shape lateral sinuses; teeth somewhat rounded; rough, bumpy surface; tufted to dense hair on lower surfaces.

shoot tips

Felty white with red margin, young leaves downy with bronze highlights.

Insect and Disease Problems

Bunch rot can be a problem due to the berries' relatively thin skin and compact clusters, but in most years with early harvest at lower sugar levels, severe rot is not common. Burger is moderately susceptible to powdery mildew. Leafhoppers can be a problem in some cases, especially as an annoyance to hand-pickers at harvest. Leafroll virus is very common in vineyards established with non-certified budwood and may cause decreased sugar accumulation and delayed harvest.

Other Cultural Characteristics

Sugar levels at maturity tend to be low, even with light crops, usually reaching 18 to 19° Brix. At moderate to high crop loads, sugar levels may not exceed 15 to 17° Brix at full maturity. Burger has low juice acidity at harvest, particularly when overcropped.

Winery Use

Light-colored, neutral wines of standard quality and low to moderate acidity are produced. These wines are used as cuveé blends in sparkling wines, in bulk wine blending, or for distillation and use in fortified wines. Future interest as a sparking wine base may be very limited due to the abundance of Chardonnay and other varieties.

—*Paul S. Verdegaal*

Cabernet Franc

Synonyms

In France, the variety is called Breton, Véron, Noir dur, Bouchy, Bouchet, Gros Bouchet, Carmenet, Grosse Vidure, Messanges rouge, and Trouchet noir. In Italy, it is known as Bordo and Cabernet frank.

Source

The variety may have been established in Bordeaux in the seventeenth century.

Description

Clusters: small to medium; cylindrical to slightly conical with shoulders; mostly well filled.

Berries: small; round, blue-black berries.

Leaves: medium; mostly 5-lobed; closed, narrow U-shaped petiolar sinus; lateral sinuses (particularly superior) often have small teeth at their base; relatively narrow, sharp teeth; rough, bumpy surface; light, tufted hair on lower surface.

Shoot tips: felty with red margin; first unfolded leaf has red-bronze highlights.

Cabernet franc is similar to Cabernet Sauvignon and Merlot but differs by smaller, compact, and mostly cylindrical clusters; in petiolar sinus; and teeth in lateral sinuses. Clusters are tighter than Cabernet Sauvignon due to greater berry set.

Growth and Soil Adaptability

Vines grow vigorously in many soil types in both cool and warm regions, thus it is generally advisable to avoid highly fertile, deep soils. Well-drained soils also help keep vigor in check. Vine spacing should be a minimum of 6 feet. Shoots grow upright, which facilitates vertical-shoot positioning. Budbreak and ripening precede that of Cabernet Sauvignon.

Rootstocks

Moderate- to low-vigor rootstocks are recommended to discourage additional vegetative growth, although several different rootstocks are in use.

Clones

Cabernet franc FPS 01 was the only registered selection available in California in the mid-1990s. This selection is reported to be high-yielding compared to clones developed in Europe. Registered and provisional selections have become available at FPS, but these have not been evaluated under California conditions. Cabernet franc FPS 03 came from Conegliano, Italy, known there as Cabernet franc ISV1. A Rauscedo selection has been registered as Cabernet franc FPS 09 (VCR10). The generic French clones available at FPS are Cabernet franc FPS 04 (French 332), 05 (French 331), 11 (French 214), 12 (French 327), and 13 (French 312). In California, ENTAV-INRA® 210, 212, 214, 327, and 623 are all available.

clusters

Small to medium; cylindrical to slightly conical with shoulders; mostly well filled.

berries

Small; round, blue-black berries.

Production

This is a moderate-yielding variety averaging 5 to 7 tons per acre. High set may require thinning at 90 percent veraison on valley floors to maximize uniformity. The variety is appropriate for hillside developments since it produces well.

Harvest

Period: A mid- to late-season variety, harvested September to November, depending on the region and crop load.

Method: Cabernet franc's moderately long bunch stem makes it easy to pick by hand. With machine harvesting, single berries, cluster parts, or entire clusters may be removed if fruit maturity is variable. Harvestability is easy to medium with canopy shaking, and juicing is light to medium. Moderately tough skins result in low juicing.

Training and Pruning

Commonly cordon trained and spur pruned, Cabernet franc may also be head trained and cane pruned. It is easily hedged and a good candidate for mechanical pruning if not cane pruned.

Trellising and Canopy Management

The fruiting wire height for a vertical-shoot-positioned system can be 30 to 42 inches high, and a 4-to 5-foot canopy above the fruiting wire is common. Two pairs of moveable wires are sufficient. In extremely vigorous sites, a horizontally divided canopy such as a lyre or "U" is used. Sunburn can occur on highly exposed fruit so a GDC is not recommended, and leaf removal is light to moderate.

leaves

Medium; mostly 5-lobed; closed, narrow U-shaped petiolar sinus; lateral sinuses (particularly superior) often have small teeth at their base; relatively narrow, sharp teeth; rough, bumpy surface; light, tufted hair on lower surface.

Insect and Disease Problems

Cabernet franc is a good indicator for leafroll virus, which is common in older plantings. On average, it is more susceptible to Pierce's disease than Cabernet Sauvignon, yet less susceptible to Eutypa dieback. Bunch rot is not a significant disease problem.

Other Cultural Characteristics

This can be a higher-yielding variety than Cabernet Sauvignon due to greater set. Late and uneven veraison is common, thus cluster thinning at this time is usually warranted to enhance ripening uniformity. Occasionally set may be reduced by shelling; however, this occurs much less severely than in Merlot or Cabernet Sauvignon.

Winery Use

As a varietal wine, it usually has a lighter body with less tannin and acid than Cabernet Sauvignon. As a result, it is often blended with this variety and occasionally with Merlot. Wines can have a pronounced vegetative aroma that is commonly associated with highly vigorous growing sites.

—*Rhonda J. Smith*

Cabernet Sauvignon

Synonyms

In France, the variety is known as Petite Cabernet, Vidure, Petite-Vidure, Bouche, Petite-Bouche, Bouchet Sauvignon, and Sauvignon Rouge. In Spain it is called Burdeos Tintos.

Source

Cabernet Sauvignon is the most important variety in the Bordeaux region of southwest France, but it is increasingly important in the Languedoc area of southern France. It is also grown widely in eastern Europe, Australia, Chile, Argentina, and, in the United States, in California and Washington. In California the variety has increased dramatically in the past 20 years in moderately warm regions, especially in high Winkler Region II to high Region III, such as central Napa Valley, as well as in Region IV, such as Lodi in the northern San Joaquin Valley. Recently Cabernet Sauvignon was shown to be a cross between Cabernet franc and Sauvignon blanc.

Description

Clusters: small to medium; conical, loose to well-filled clusters; medium-long peduncles.

Berries: small; round, blue-black berries; thick skins.

Leaves: medium; very deeply 5-lobed; overlapping, lyre-shaped petiolar sinus and lateral sinuses that appear like five round holes around leaf margin; medium-sized teeth; upper surface dark green and smooth; scattered tufts of hair on lower surface.

Shoot tips: felty with red margins; young leaves with bronze-red cast.

Growth and Soil Adaptability

The vine is vigorous to excessively vigorous depending on the site, rootstock, planting density, and trellis system interactions; it is rarely inadequately vigorous. Shoots are initially upright but trail in late season when growth is abundant. Production of a second crop can be moderate when vigor is excessive. Recommended vine spacing is 5 to 7 feet depending on site conditions. High-density plantings (1,500 vines per acre) can cause uncontrollable growth. Regardless of rootstock choice, Cabernet Sauvignon does not perform well on poorly drained soils.

Rootstocks

Rootstocks are usually selected to counteract scion vigor. On deep soils with high-vigor potential, the low-vigor rootstocks 3309C, 101-14 Mgt, or 1616C are preferred; on moderately deep soils Teleki 5C or SO4; and on shallow soils or hillsides where irrigation is limited the drought-tolerant and vigorous rootstocks 110R and 140Ru are preferred. The rootstock 420A can only be recommended on deep, fine-textured soils and at relatively close spacing. There is interest in Riparia Gloire for deep, fertile soils but relatively little data or experience is available. While St. George has been a traditional combination for Cabernet Sauvignon, soils must not be root restricted, and a tendency for poorer fruit set exacerbates excessive growth.

clusters

Small to medium; conical, loose to well-filled clusters; medium-long peduncles.

berries

Small; round, blue-black berries; thick skins.

Clones

The most widely planted Cabernet Sauvignon clones are the identically performing selections Cabernet Sauvignon FPS 07 and 08. (Both originated from the same Concannon vines, differing only in days of heat treatment.) Cabernet Sauvignon FPS 06 is a highly regarded selection from the former UC Jackson Foothill Experiment Station; however, its yield is about 60 percent that of FPS selection 08. Whether the trade-off in yield is balanced by a commensurate increase in wine quality is a matter of debate. Interest has also surfaced in Cabernet Sauvignon FPS 04 from Mendoza. More recently, new clonal material is available for which there is little California performance information. From France, Cabernet Sauvignon ENTAV-INRA® 15, 169, 170, 191, 337, 338, 341, and 412 should increase the diversity of Cabernet materials available to growers and vintners. Viticulturists should be wary of traditional field selections of Cabernet Sauvignon. Many are infected with leafroll and/or corky bark viruses and exhibit graft incompatibilities with certain rootstocks. Three of these field selections from the Napa Valley have been treated for virus and are now available as registered selections: Cabernet Sauvignon FPS 29 (Niebaum-Coppola Vineyards), Cabernet Sauvignon FPS 30 (Disney Silverado Vineyards), and Cabernet Sauvignon FPS 31 (Mondavi Vineyards).

Production

Vineyards can bear 6 to 7 tons per acre, especially at higher planting densities, but lower yields (3 to 4 tons) are expected when grown on hillsides or shallow soils. The crop is often thinned significantly at veraison (up to 20 percent) to eliminate later-ripening fruit.

Harvest

Period: A mid- to late-season variety; harvest occurs in late August to late October, depending on the location and heat unit accumulation.

Method: Hand harvest is the typical method for the variety in high-quality regions. Machine harvest with a canopy shaker is easy to medium, with fruit removed mostly as single berries and some clusters. Juicing is light. Trunk shaking is easy, with mostly single berries and some clusters removed. Juicing is also light. There is less MOG and spur damage with trunk shaking when compared to straight-rod heads. Bow-rod machines picking vines trained to vertical-shoot-positioned systems will have less MOG than unpositioned trellis systems.

leaves

Medium; very deeply 5-lobed; overlapping, lyre-shaped petiolar sinus and lateral sinuses that appear like five round holes around leaf margin; medium-sized teeth; upper surface dark green and smooth; scattered tufts of hair on lower surface.

shoot tips

Felty with red margins; young leaves with bronze-red cast.

Training and Pruning

Vines are commonly trained to a bilateral cordon and pruned to six to eight two-node spurs per 3 feet of cordon. Unilateral cordons may be used when vines are spaced less than 5 feet apart. Quadrilateral cordon training is often practiced where high vigor is expected. There are strong proponents of cane pruning the variety in the North Coast. Cabernet Sauvignon is very adaptable to mechanical pruning and harvest.

Trellising and Canopy Management

Vertical-shoot-positioned systems are generally used whether as undivided or as horizontally divided systems. Vertically divided systems such as Scott Henry or Smart-Dyson are sometimes used to balance expected vigor with additional retained nodes in the fruiting zone. GDC systems are falling out of favor as the upright growth of Cabernet Sauvignon is difficult to position in a downward manner. Orienting rows of VSP systems to prevent direct exposure of fruit to afternoon sun is desirable. Trimming shoot growth in VSP systems is common, but a trellis should provide approximately 4 feet of canopy after trimming for proper balance of fruit and leaf area.

Insect and Disease Problems

Cabernet Sauvignon is moderately susceptible to powdery mildew. Its loose clusters are not prone to bunch rots, but rot can occur when spur selection at pruning results in crowded shoot growth. The variety is quite susceptible to Eutypa dieback; vineyards 12 to 15 years old can have a high percentage of vines with at least one point of infection. Vineyards older than 15 years often also exhibit symptoms of black measles. Illegally imported clones are strongly suspected to be infected with viruses, probably leafroll or corky bark, or a mixture both.

Other Cultural Characteristics

Cabernet Sauvignon often produces blind buds, typically in the mid-cane region, nodes 6 through 10 on a 12-node cane. This is a particular problem in the training phase when long canes are laid out on the fruiting wire to become cordons. Blind buds on those canes preclude normal spacing of spurs. Latent shoot production is prodigious, and shoot thinning removes unwanted shoots at the base of two-node spurs or at the head of cane-pruned vines.

Winery Use

Cabernet Sauvignon is used exclusively for high-quality to middle-quality, dry table wines. Small amounts may be blended into Merlot or Cabernet franc wines to provide more tannin structure. Grapes grown in areas too cool for the variety can develop a highly undesirable herbaceous or "green bell pepper" aroma in very shaded conditions. In areas too warm for the variety, fruit will not develop normal varietal character.

—James A. Wolpert

Carignane

Synonyms
In Spain the cultivar is referred to as Cariñena and Mazuelo, while in France it is known as Carignan noir and Monestel. In Portugal it is called Pinot Evara, while in Italy it is known as Carignano.

Source
The variety is originally from the northeastern Spanish province of Aragon, near the town of Cariñena. Known in France since the mid-twelfth century, Carignane was originally planted in the Pyrenees Orientales. From there plantings spread throughout the Midi region, where it is used for the production of common red table wine. It remains the most cultivated grape variety in southern France.

Description
Clusters: medium to large; broad-conical, well-filled to compact clusters; medium-long, well-lignified peduncle.

Berries: medium; short oval; dark purple-black with a gray bloom and thick skin.

Leaves: large; moderately 5-lobed, closed U-shaped petiolar sinus and narrow lateral sinuses; leaf tissue does not lie flat near the petiole and puckers up into a "target patch" where the main veins join the petiole; relatively large, broad, sharp teeth in several ranks; glabrous to sparse tufted hair on lower leaf surface.

Shoot tips: felty white, slight red margin; young leaves yellowish green.

Growth and Soil Adaptability
Own-rooted vines grow vigorously on fertile, medium-textured to heavy soils. Due to potentially high vigor, the variety is also well adapted to hillsides or sites with limited soil depth or fertility in coastal regions. Growth is upright to semi-erect and often open, with large canes. Canes harden off early in the season and mature well. The recommended in-row spacing for bilateral cordon vines is 7 to 8 feet in the San Joaquin Valley and 6 feet in coastal regions.

Rootstocks
Historically, Carignane was planted on its own roots in the San Joaquin Valley and on Rupestris St. George rootstock on hillside plantings in coastal regions. In coastal regions where phylloxera resistance is desired, moderate-vigor rootstocks such as 101-14 Mgt, SO4, Kober 5BB, and 3309C may be used. In hillside plantings, or in areas where soil depth or fertility is limited, 110R, 1103P, and 140Ru are acceptable choices. In the San Joaquin Valley, where nematode resistance is desired, Freedom, Harmony, and 1103P may be likely choices.

Clones
Carignane FPS 02 and 03 are currently available as registered stock. They are subclones of Carignane FPS 01 and could be expected to perform identically. Commonly cultivated clones in California include selections 01 (non-registered) and 02; both have acceptable fruiting characteristics. Carignan

clusters
Medium to large; broad-conical, well-filled to compact clusters; medium-long, well-lignified peduncle.

berries
Medium; short oval; dark purple-black with a gray bloom and thick skin.

ENTAV-INRA® 6 is now available in California. (There is no "e" in the French spelling). Unfortunately, information regarding viticultural characteristics or relative performance of these Carignane selections is not available.

Production

Carignane is a highly productive variety, with yields ranging from 10 to 14 tons per acre in the San Joaquin Valley, and 4 to 8 tons per acre in coastal regions.

Harvest

Period: A late-season variety, typically ripening in late September to mid-October in the San Joaquin Valley, Carignane may be harvested earlier in this region if used for blush wine production. Harvested in mid- to late October in coastal regions.

Method: Clusters have moderately thick and short peduncles, which require that knives or shears be used when hand-harvesting fruit. Canopy shaking makes harvesting difficult, with fruit removed as single berries. Considerable force is needed to remove fruit, resulting in moderate to heavy juicing, excessive defoliation, and damage to spurs and canes. Vines are more difficult to harvest when soluble solids reach 23 to 24° Brix. Trunk shaking results in intermediate to difficult harvesting with medium juicing. Fruit is removed mostly as single berries and a few cluster parts.

Training and Pruning

Carignane is commonly trained to bilateral cordons and spur pruned, retaining 12 to 16 two- to three-node spurs per vine. Many older vineyards in both the San Joaquin Valley and coastal regions were head trained and spur pruned. Twelve to 14 two- to three-node spurs are typically retained on head-trained vines. Fruit from head-trained vines is often preferred for the fresh juice grape market.

Trellising and Canopy Management

Vines grown in coastal regions may be head trained or trellised to vertical-shoot-positioned systems. In the San Joaquin Valley vines are

leaves

Large; moderately 5-lobed, closed U-shaped petiolar sinus and narrow lateral sinuses; leaf tissue does not lie flat near the petiole and puckers up into a "target patch" where the main veins join the petiole; relatively large, broad, sharp teeth in several ranks; glabrous to sparse tufted hair on lower leaf surface.

shoot tips

Felty white, slight red margin; young leaves yellowish green.

generally trellised to the traditional California two-wire system, although some head training is still used. Basal leaf removal may be performed to improve canopy microclimate and increase spray penetration into the fruiting zone.

Insect and Disease Problems

Both foliage and berries are extremely susceptible to powdery mildew, especially bud penetration. A thorough mildew control program, including the use of sulfur and sterol-inhibiting fungicides, is recommended to prevent infection and manage resistance. Compact clusters may result in summer bunch rot complex in the San Joaquin Valley. Carignane is also sensitive to downy mildew and Eutypa dieback canker, and it is moderately susceptible to Botrytis bunch rot following rains near harvest. Susceptibility to both summer bunch rot and Botrytis is accentuated by fruit damage from powdery mildew infections. Basal leaf removal may be performed to increase fungicide penetration and reduce humidity in fruiting zone.

Many older vineyards may be infected with leafroll and fanleaf viruses, causing reductions in yield and overall fruit quality. Some older plantings also contain vines infected with tomato ringspot virus, commonly referred to as yellow vein virus. Infected vines produce excessively loose, poorly filled clusters with shot berries. This results in severe yield reductions and increased vine vigor or size, thus the expression "unfruitful Carignane" is often used to describe infected vines. For these reasons care must be taken to propagate only virus-free, certified planting stock.

Other Cultural Characteristics

Budbreak is typically late, and the variety may not fully ripen in cool, coastal regions.

Care should be taken to avoid overcropping, which results in low sugar and acidity, as well as poor color development. Carignane produces a significant second crop, which ripens several weeks after the primary crop. Second-crop removal or selected hand harvest may be necessary in coastal regions.

Winery Use

In the San Joaquin Valley, Carignane produces standard red table or blending wines with moderate to good color and significant tannin but little pronounced varietal flavor. In some cases the wine may be bitter or harsh. It is also used for the production of rosé or blush wines in the San Joaquin Valley. In coastal regions, Carignane produces a slightly more complex, varietal table wine and may also be used in Rhône-style blends. Carignane is also an important fresh juice grape variety for home winemaking.

—*Nick K. Dokoozlian*

Chardonnay

Synonyms

Chardonnay has been referred to as Pinot Chardonnay in California. The variety has often been confused or associated with Pinot blanc and Melon in many wine regions of the world. However, the three varieties are distinct from each other and can be separated by morphological differences and more recently by genetic comparative testing. Recent genetic studies indicate that Chardonnay is a cross between Pinot noir and Gouais blanc.

Source

Chardonnay is one of the leading white grape varieties in the world for production of high-quality white wines. The variety probably originated from the Burgundy region of France. The name "Chardonnay" has been linked to a village in the Mâcon region and is derived from Cardonnacum, which means "a place of chardons or thistles." In California there are references to Chardonnay being grown in the late 1800s. Plantings remained limited due to Chardonnay's low fruit yields compared to the higher-yielding varieties grown at that time. During Prohibition most Chardonnay vineyards were uprooted in favor of varieties that could withstand shipment to the East Coast. After Prohibition two surviving Chardonnay vineyards were the Wente Vineyard in Livermore and Paul Masson's La Cresta Vineyard in the Santa Cruz Mountains. These vineyards are believed to represent different introductions of Chardonnay into California. Budwood collected from the Wente Vineyard has been a major source for the expansion of Chardonnay acreage in the state. In 1960 it was estimated that only 150 acres of Chardonnay existed in California. By 2000 there were 103,491 acres reported, making it the state's most widely planted wine grape variety.

Description

Clusters: small to medium; cylindrical, often winged to double in larger clones, short peduncles. French clones are typically smaller cylindrical clusters; they are often larger in California.

Berries: small; round, yellow to amber when ripe.

Leaves: medium; more or less entire with shallow lateral sinuses; U-shaped petiolar sinus with naked veins; short, broad teeth; upper surface bullate and rough; lower surface mostly glabrous with scattered hairs.

Shoot tips: downy white; young leaves yellow-green with subtle bronze-red tinges.

Growth and Soil Adaptability

Vine vegetative growth can vary significantly from weak to moderately vigorous depending on climatic region, soil, virus status, and rootstock selection. Adaptable to a wide range of soil types, Chardonnay's highest vigor will be on deep valley bottom soils

clusters

Small to medium; cylindrical, often winged to double in larger clones, short peduncles. French clones are typically smaller cylindrical clusters; they are often larger in California.

berries

Small; round, yellow to amber when ripe.

with high moisture availability. Growth on low-vigor sites can be influenced by rootstock selection. Using higher-vigor rootstocks can improve vine growth during vineyard establishment. In the cooler coastal areas, persistent winds can significantly reduce the growth and yield capacity of Chardonnay vines. Shoots in non-positioned canopies will display a trailing growth habit, especially when vines have high vigor. Planting density will depend on potential vine vigor, spacing, and trellis design. Vine in-row spacing can vary from 4 to 6 feet apart.

Rootstocks

Chardonnay has no known incompatibilities when FPS-certified budwood is used to propagate planting stock. Field selections that are infected with viruses can cause graft incompatibilities. Chardonnay has been grown successfully on a wide variety of rootstocks. Rootstock selection should be based on the pest situation, soil characteristics, and potential vine vigor of the site. In low-vigor sites the use of higher-vigor rootstock varieties can reduce the time needed to train vines. When nematodes are present, the selection is more limited due to the lack of nematode tolerance of many of the rootstocks. In the Central Coast, rootstocks have been observed to influence the severity of winter injury on 1- to 3-year-old Chardonnay vines in areas where temperatures during the winter commonly fall well below 32°F. The rootstocks 110R and 3309C appear to have more severe symptoms than other selections.

leaves

Medium; more or less entire with shallow lateral sinuses; U-shaped petiolar sinus with naked veins; short, broad teeth; upper surface bullate and rough; lower surface mostly glabrous with scattered hairs.

Clones

In the early 1950s, Louis P. Martini made selections from the Stony Hill Vineyard in the Napa Valley, which had been planted as a mass selection of budwood from the Wente Vineyard. An early University of California clonal selection effort in the late 1950s by H.P. Olmo identified more productive clones. These early clonal studies demonstrated the possible yield improvement for a variety that had had limited interest due to low-yielding selections. Chardonnay FPS selections 04 to 14 were eventually released from the Martini material. FPS selections 04 and 05 are the most widely planted in California. Some confusion remains over the term "Wente clone." The name has described both an older selection with small clusters that have a high percent of shot berries (often called "old Wente") and the more productive heat-treated selections from FPS that can be traced back to the Wente Vineyard.

A number of field selections have also been propagated from older vineyards with a history of high wine quality. Budwood from the Mount Eden Vineyard in the Santa Cruz Mountains, which was planted from budwood from Paul Masson's La Cresta Vineyard, most likely represents a separate introduction into California. It is a low-yielding, virus-infected selection with small berries and tight clusters. There are also Chardonnay musqué selections such as Rued, See's, Spring Mountain, and Sterling (FPS 79 and 80) that have a slight muscat-like character.

Nearly one hundred clones and subclones of Chardonnay are now available in California. This is due to the aggressive addition of California heritage selections of Chardonnay to the registered Foundation Vineyard collection (including those mentioned above), combined with importation from Italy and especially, France. Available in the ENTAV-INRA® trademark program are clones 76, 95, 96 124, 131, 277, 548, and 809. Just a sample of the generic certified selections includes Chardonnay FPS 37 (French 95), FPS 39 (French 78), FPS 40 (French 125), FPS 41 (French 352), FPS 42 (French 277), FPS 69 (French 76), FPS 70 (French 96), and FPS 81 (French 117).

Recent clonal evaluations among Chardonnay selections have shown differences in yield, vigor, fruit intensity, and flavor profiles. There is no one best selection; the most complexity may be achieved by blending wines made from several selections. Wines produced from a single clone can vary greatly due to climatic region and site conditions.

Production

Vine yield can vary considerably by climatic region, clonal variation, and cultural practices. Crop size can range from 2 to 8 tons per acre. The variety was originally grown mainly in the coastal production areas; significant acreage now exists in the warmer interior valleys.

Harvest

Period: An early season variety, in warmer regions ripening in late August to early September; and in colder production areas ripening mid- to late October.

Method: Short peduncles and many small bunches slow hand harvest, but cluster stems are thin and easy to cut with knives or shears. Harvest is easy to moderately easy with horizontal-rod or bow striker machines. Fruit comes off mostly as single berries with moderate juicing. Bow rods picking well-trained vines on vertical-shoot-positioned trellises have lower shoot and spur breakage than straight-rod heads. Shoot breakage can be high due to the brittleness of the wood, especially with dense foliage. Harvestability is medium with trunk shakers. Fruit comes off as single berries with medium juicing. Adding straight rods to the picking head can improve crop removal in some vineyards.

Training and Pruning

Vines are commonly trained to bilateral cordons and spur pruned. On low-vigor sites, higher plant densities and the use of unilateral cordons may produce vines that better balance fruit and vegetative growth. Quadrilateral cordon training should be used only on high-vigor sites. The number of spurs left per vine will depend on in-row vine spacing, bud fruitfulness, and vines' capacity for yield. For selections with small clusters or when bud fruitfulness is low, the use of head training or short cordons with cane pruning may improve vine yields. Chardonnay is adaptable to mechanical pruning systems. In cooler production areas, crop levels need to be managed to avoid overcropping, which can result in delay of fruit ripening and potential loss in shoot vigor.

Trellising and Canopy Management

For low- to moderate-vigor sites, vertical-shoot-positioned systems are appropriate. For higher-vigor sites, horizontal or vertical splitting of the canopy can improve canopy and fruit exposure and reduce canopy shade. Lyre and GDC systems are two of the horizontally split systems. GDC systems may result in higher fruit exposure and increase the potential for sunburn in warmer production areas. Smart-Dyson and Scott Henry are vertical systems that can be used to reduce canopy density on higher-vigor sites.

Insect and Disease Problems

Chardonnay is highly susceptible to powdery mildew and Pierce's disease. Botrytis bunch rot susceptibility varies among selections: those with tight clusters are more prone to rot especially when preharvest rains occur. Leafroll virus and virus-associated graft incompatibilities are potential problems in some field selections. The use of certified planting stock is highly recommended.

Downy white; young leaves yellow-green with subtle bronze-red tinges.

Other Cultural Characteristics

Chardonnay is one of the first varieties to begin growth in the spring, which makes vines more susceptible to frost injury. Crop recovery from regrowth is usually small. In years when cool bloom temperatures occur a high percentage of seedless, shot berries can occur. Early maturity makes the fruit attractive to birds, and significant fruit loss can occur.

Winery Use

A wide range of wine styles can be produced, including sparkling wines. Wine characters are greatly influenced by fermentation variables and the use of oak in the winery. Climatic conditions, soil characteristics, and viticultural farming practices are the major factors influencing vine growth and potential wine quality. The most distinctive wines are produced in areas where the grapes can achieve full ripeness and yet maintain moderately high acidity. As in most growing regions of the world, grapes for the highest-quality Chardonnay wines are grown in the cooler climatic regions.

—Larry J. Bettiga

Chenin Blanc

Synonyms
Chenin is the official name in France, where it is commonly called Pineau de la Loire, and, less often, Pineau d'Anjou. In South Africa it is commonly called Steen.

Source
Chenin blanc is an old variety from Anjou, France, known to have been growing there since 845 AD and then spreading to neighboring areas. It is a leading variety of the middle Loire region where it is used to produce dry and natural sweet table wines as well as sparkling wines. It did not emerge in California until after World War II when some North Coast wineries acquired vines from the collection at UC Davis for premium table wine production. Acreage expanded rapidly in the 1970s, peaked in the early 1980s, and has since declined. Now it is mostly grown in moderately warm coastal valleys, the Sacramento and San Joaquin Delta, and the San Joaquin Valley for the production of table and sparkling wines.

Description
Clusters: medium to large; long conical, compact, often winged; short to medium peduncles.
Berries: medium; oval, yellow-green.
Leaves: medium; 3- to 5-lobed with U-shaped petiolar sinus; inferior lateral sinuses often shallow; short teeth; moderately dense hair on lower leaf surface; leaf veins near the petiolar junction pink-red and noticeable on upper surface.
Shoot tips: felty white; dense hair on young leaves makes them appear cream-white.

clusters
Medium to large; long conical, compact, often winged; short to medium peduncles.

berries
Medium; oval, yellow-green.

Growth and Soil Adaptability
Vines are very vigorous when grown on their own roots in medium- to fine-textured soils (sandy loam to clay loam); they show poor vigor on very sandy soils. Vines are more vigorous than Chardonnay but less vigorous than Sauvignon blanc and Colombard. The vines leaf out early and have a spreading growth habit. Recommended in-row spacing is 6 feet in poor soils or in coastal regions and 7 feet in good soils or in Central Valley regions.

Rootstocks
Chenin blanc has no known incompatibilities. Freedom and Harmony rootstocks are used in the San Joaquin Valley for nematode resistance. Since the failure of AXR #1, experience with phylloxera-resistant rootstocks in California is very limited.

Clones
Registered selections in California have been limited to selections from regional commercial vineyards. Chenin blanc FPS 02, 03, and 04 were derived from FPS 01 using heat therapy. FPS 05 was established from a different California vineyard. A comparative trial demonstrated that Chenin blanc FPS 04 was the most productive, followed closely by selection 01. Selection 05 should not be planted because of its higher bunch rot potential (75 percent increase over FPS Chenin blanc 01 and 04), which is due to small, very compact clusters—in spite of its smaller berries. No differences were shown in sensory analysis of experimental wine lots

from these clonal trials. The discovery of a loose-clustered, virus-free clone would benefit this variety.

Production

A consistent producer, Chenin blanc usually yields 8 to 11 tons in the Central Valley and 5 to 8 tons per acre in coastal regions.

Harvest

Period: Chenin blanc is a midseason variety, but harvest is practiced early (mid-August to mid-September) in the San Joaquin Valley due to a greater bunch rot potential. Harvest is mid-September to mid-October in the coastal valleys.

Method: Clusters have thick, short to medium-long peduncles, which require the use of knives or shears to hand harvest. Canopy shakers result in medium harvestability and medium juicing. The fruit is mostly removed as single berries with some cluster parts. Trunk shakers are considered the best harvesting method for this variety. They result in easy to medium harvestability, medium juicing, but less than with rod-type canopy shaker. Fruit is removed as single berries and cluster parts as well as some whole clusters. Fewer rotten clusters are removed with trunk shakers.

Training and Pruning

Chenin blanc is mostly trained to a bilateral cordon and pruned to 12 to 16 two-node spurs. Additional numbers of spurs may be needed for large vines and to minimize tight clusters by increasing cluster numbers. Cane pruning reduces bunch rot potential with less compact clusters but increases the cost of pruning and tying.

The variety is quite fruitful and the closely spaced nodes on spurs easily contribute to high node numbers at pruning. This characteristic makes it less suitable to machine hedge pruning that can delay fruit maturation, the result of overcropping.

leaves

Medium; 3- to 5-lobed with U-shaped petiolar sinus; inferior lateral sinuses often shallow; short teeth; moderately dense hair on lower leaf surface; leaf veins near the petiolar junction pink-red and noticeable on upper surface.

shoot tips

Felty white; dense hair on young leaves makes them appear cream-white.

Trellising and Canopy Management

Leaf removal from the cluster region after veraison may reduce bunch rot potential. Cluster exposure may be facilitated with vertical-shoot-positioned systems when using single and divided canopy systems.

Insect and Disease Problems

Tight clusters can contribute to bunch rot. Problems may be minimized by early deficit irrigation before harvest, increasing cluster numbers by retaining more buds at pruning, and cluster exposure with leaf removal. Pre-bloom gibberellin "stretch" spray application to elongate the cluster stem structure has been practiced by growers to reduce bunch rot. The practice is currently limited to those receiving a Special Local Need label permit from California Department of Food and Agriculture. Growers should be aware that excessive gibberellin rates or improper application can reduce current season's yield as well as return bud fruitfulness the following season. Bloom-time and pre-bunch-closure fungicide sprays may help reduce Botrytis bunch rot. In some years during cool weather, large numbers of flower thrips may retard early shoot growth. The variety is somewhat tolerant of Pierce's disease and very susceptible to Eutypa dieback.

Other Cultural Characteristics

Three- or four-year-old vines tend to over-produce; shoot and cluster thinning at this age is often needed. Shoots adhere firmly and are not easily blown off by high winds in the spring, a reason for this variety's popularity in South Africa. Chenin blanc has moderately good acid level, attaining its best balance of sugar and acid in the cool to moderately warm growing regions.

Winery Use

Chenin blanc is used to produce quality, well-balanced table wines, usually under a varietal label. Coastal wines are usually moderately distinct and fruity in character. It produces some very good sparkling wines and can be used for natural sweet wines in cool districts.

—*L. Peter Christensen*

Colombard

Synonyms

Colombard is the official name in France and the most widely used name worldwide. It is also called Colombar and Colombier in France, French Colombard in California, and Colombar in South Africa.

Source

The variety was first cultivated in southwest France and is now mainly used for brandy, including Cognac and Armagnac. It is also used as a supplementary variety in the white wines of Bordeaux and other districts. First brought to California in the 1850s, Colombard was grown for years in the Central Valley as "West's White Prolific," after George West, a prominent San Joaquin County wine grape producer. Acreage was very limited until the table wine boom in the 1960s–70s when it became the state's most widely planted variety. Plantings peaked at 90,000 acres in 1987, with the greatest concentrations in the San Joaquin Valley. The California plantings are still the largest in the world, followed by South Africa.

Description

Clusters: medium; long conical to cylindrical, well-filled, often winged or double; long peduncles.
Berries: medium; round to short oval; yellow-green with high acid levels when ripe.
Leaves: Large to medium-large, mostly 3-lobed to almost entire; wide V-shaped petiolar sinus; short, sharp teeth; moderately dense hair on lower leaf surface; leaves can be pinched; petioles pink-red with green veins.
Shoot tips: felty white; young leaves yellow and downy.

Growth and Soil Adaptability

Own-rooted vines are exceptionally vigorous and develop large canopies when grown on medium- to fine-textured soils (sandy loam to clay loam); they are sometimes excessively vigorous on very fertile soils, and show moderate to poor vigor on very sandy soils, largely due to the variety's high susceptibility to root knot nematodes. Colombard is more tolerant of saline and high boron soil conditions than many other varieties, probably the result of dilution by large canopies. Recommended in-row spacing is 7 to 8 feet; 8 feet is preferred in good soil sites. Row spacing of less than 10 feet in vigorous sites may be too crowded for conventional equipment due to large canopies.

Rootstocks

Own-rooted Colombard vines are highly susceptible to root knot nematodes in sandy soils. Freedom, Ramsey, and Harmony successfully avoid this problem but may be too vigorous in fertile, medium- to fine-textured soils. Phylloxera-resistant rootstocks should be selected for moderate vine vigor characteristics in fertile soils to avoid excessive vigor problems associated with Colombard.

clusters

Medium; long conical to cylindrical, well-filled, often winged or double; long peduncles.

berries

Medium; round to short oval; yellow-green with high acid levels when ripe.

Clones

Available registered selections in California have been limited to selections from regional commercial vineyards. The clonal differences measured in comparative trials have been minor. Yield component, fruit composition, and wine sensory analyses of French Colombard FPS 01, 02, and 05 over six years have shown too little difference to demonstrate clonal preferences. Selection 02, which has been widely available through the California grapevine nursery industry, shows good vine and fruit characteristics.

Production

Colombard usually yields 8 to 13 tons per acre. Higher yields are commonly achieved in young, vigorous vineyards and those trained to quadrilateral systems.

Harvest

Period: In the San Joaquin Valley, Colombard is a midseason variety, with harvest beginning in Kern County in mid- to late August and ending in mid- to late September in Lodi. In the North Coast, Colombard is usually the last white variety to be harvested. Harvest can be substantially delayed by heavy cropping.

Method: The relatively thick cluster stem, which is short to medium in length, must be cut for hand harvest. Dense foliage can interfere with harvesting. Pre-harvest cane trimming may be helpful on trellised vines. Canopy shakers result in medium harvestability, with fruit mainly removed as single berries and some cluster parts. The wood is brittle, and machine harvesting can cause some spur breakage. Pre-harvest trimming can decrease the interference by dense foliage and strong shoot growth. Trunk shakers result in easy to medium harvestability and medium juicing. Fruit is mostly removed as single berries and some cluster parts. Trunk shakers cause less cane breakage than do canopy shakers with rods.

Training and Pruning

Vines are most commonly trained to bilateral cordons with spur pruning. Blind buds on young cordons of vigorous vines are a problem during vine training. Cordon branch canes (usually trained during second leaf) exceeding ⅝ inch diameter are most susceptible to poor emergence of dormant buds. A solution is always to prune the lateral shoots to one-node spurs at each intended spur position. Training vines of moderate vigor also helps. This can be accomplished by training in the first year of vine establishment. For training during second leaf, leave four or more lateral shoots on the vine trunk to dilute vine vigor while the cordons are being trained.

leaves

Large to medium-large, mostly 3-lobed to almost entire; wide V-shaped petiolar sinus; short, sharp teeth; moderately dense hair on lower leaf surface; leaves can be pinched; petioles pink-red with green veins.

shoot tips

Felty white; young leaves yellow and downy.

Colombard is commonly pruned to 14 to 20 spurs, although higher node numbers are sometimes used on extremely vigorous vines. Yields are usually increased with machine-hedge pruning, with a slight or no delay in fruit ripening and no change in fruit composition at harvest. The method retains all nodes within the box configuration of the spur zone. Quadrilateral cordon training with 30 spurs has successfully spread the vine canopy and fruiting zone in highly vigorous vineyards, resulting in higher yields with minimal delays in fruit ripening. Quadrilateral cordon separation of 24 to 30 inches and without foliar wires is recommended to facilitate machine harvest. Minimal pruning (no pruning except for canopy bottom) is another acceptable management system (demonstrated by experimental and commercial experience), but expect two-week delays of fruit maturation.

Trellising and Canopy Management

Vigorously growing shoots are subject to wind breakage in early spring. A foliar catch wire with bilateral cordons will reduce damage. Colombard's high vigor and large canopies respond favorably to horizontal quadrilateral cordon systems such as GDC, but they are not suited to vertical-shoot-positioned systems.

Insect and Disease Problems

Some plantings have occasionally shown "spindle shoot" symptoms on individual vines in the spring; leaves are small, yellowish, and puckered around the margins. Affected vines should not be used as a source of propagating wood due to a suspected virus-disease presence. The disorder has not appeared in certified wood sources. Shoots may be stunted by feeding western flower thrips in cool springs or grape thrips in the summer.

Bunch rot may be a problem, especially in 3- and 4-year-old vines. Colombard's dense foliage, which interferes with spray and dust coverage, results in a higher potential for powdery mildew. It is very susceptible to Phomopsis cane and leaf spot and moderately susceptible to Eutypa dieback.

Other Cultural Characteristics

In early summer, sudden heat spells may cause shoot tips to die back and occasionally damage developing clusters. Colombard is medium-late in fruit ripening, although the fruit holds well on the vine until rainfall.

Winery Use

In warm to hot climates, Colombard is a versatile variety of high productivity and good fruit composition (high acidity and low pH). It produces a fruity, crisp wine in cool districts and has sufficient acid for a balanced and distinct varietal wine or for use in blends in warm districts. Colombard is used widely in the San Joaquin Valley as a blending base of white table and sparkling wines and in the production of grape juice concentrate and brandy.

—*L. Peter Christensen*

Durif

Synonyms

The variety is known as Dure, Duret, Plant Durif, Pinot de Romans, Pinot de l'Hermitage, Plant Fourchu, Nerin, Gros Noir, and Bas Plant in France and, in California, Petite Sirah.

Source

A French nurseryman by the name of Durif first propagated the variety in the 1880s in the Rhône Valley. Recent DNA research at University of California, Davis, indicates Durif resulted from a cross between the Rhône varieties Syrah and Peloursin. Charles McIver, founder of Linda Vista Winery near Mission San Jose, was the first to import the vine into California in 1884, along with other French varieties. He and others were soon calling it Petite Syrah, and the name stuck. Older California "Petite Sirah" vineyards are often mixed plantings containing mostly Durif, but they also include varieties such as Barbera, Carignane, Peloursin, Syrah, and Zinfandel.

Durif was popular in the Central Valley during the planting boom of the 1970s, mainly to add color and tannin to generic wines. Most of these vines have been removed due to virus presence, disappointing yields, and susceptibility to sunburn and berry shrivel. Durif has regained some popularity as a niche variety in coastal plantings.

Description

Clusters: medium; long conical to cylindrical, compact, often winged to double; short to medium peduncles.

Berries: medium; short oval to round; blue-black with a silvery bloom.

Leaves: small to medium; deeply 3-lobed (shallow inferior lateral sinuses); closed U-shaped petiolar sinus; short, sharp teeth; lower leaf surface glabrous.

Shoot tips: downy, white tips with rose margin; young leaves with cobwebby hair and bronze-red highlights.

Growth and Soil Adaptability

Vines are moderately vigorous and relatively weeping in growth; fasciated shoots are common. Durif leafs out late in the spring and shoots develop slowly. Well-drained soils with moderate vigor potential are preferred for the production of high-quality wines. Because of Durif's potential for sunburn and berry shrivel, soils that readily contribute to stressed vines should be avoided, especially in warm climates.

Rootstocks

Rootstock selection should be based on soil conditions and planting density. When planted on poor soils, a vigorous rootstock such as St. George, Freedom, or 110R would be appropriate. On more fertile sites, moderate-vigor stocks such as 101-14 Mgt or 3309C can be used to limit growth and improve fruit quality.

clusters

Medium; long conical to cylindrical, compact, often winged to double; short to medium peduncles.

berries

Medium; short oval to round; blue-black with a silvery bloom.

Clones

Little, if any, clonal research has been done on this variety. The only selection currently registered is listed as Petite Sirah FPS 03. An old Napa Valley selection is currently in the virus testing and virus therapy process.

Production

Durif vines are moderately productive with medium, compact clusters. Yields range from 3 to 4 tons per acre in coastal and foothill counties and 5 to 8 tons in the Central Valley.

Harvest

Period: Durif is a mid- to late-season variety. In cooler regions, harvest may not occur until October.

Method: Hand harvest is easy with knives or shears. Moderate growth makes the fruit accessible. Machine harvest with canopy shakers is medium, with juicing medium to heavy. Trunk shaking is medium-hard, with medium juicing. Overripe fruit is difficult to remove.

Training and Pruning

Traditionally, Durif vineyards were head trained and spur pruned in a similar fashion to Zinfandel. Some wine producers have planted new vineyards in this same fashion. Head-trained vines on shallow soils should be limited to 7 to 10 spurs per vine. Ten to 12 spurs per vine are satisfactory with bilateral cordon-trained vines in deeper coastal soils; 12 to 14 spurs per vine may be used in warmer districts.

leaves

Small to medium; deeply 3-lobed (shallow inferior lateral sinuses); closed U-shaped petiolar sinus; short, sharp teeth; lower leaf surface glabrous.

Downy, white tips with rose margin; young leaves with cobwebby hair and bronze-red highlights.

Trellising and Canopy Management

Unless vines are head trained, vertical-shoot-positioned systems are appropriate. The moderate growth usually requires a minimal amount of canopy manipulation. Split canopy systems should be considered only on sites with especially high potential vigor. Leaf removal in the fruit zone is useful to reduce the risk of Botrytis bunch rot; it should be avoided on south and west canopy exposures due to the potential for sunburn.

Insect and Disease Problems

Bunch rot can be a problem due to Durif's compact bunches and late ripening period. Durif is considered fairly tolerant of powdery mildew. Leafroll and corky bark virus diseases were once prevalent in older plantings. Certified planting stock should be used.

Other Cultural Characteristics

The fruit tends to sunburn and raisin if the vines are stressed for moisture during hot spells, or if the fruit is exposed. Upon maturity, the fruit begins to shrivel and raisin, especially in warm climates. Therefore, timely harvest is important to minimize fruit weight loss and to maintain quality. The vines may produce a substantial second crop in some years. Durif has fairly good crop recovery following spring frost damage.

Winery Use

Durif produces a full-bodied, red table wine with deep color and long aging potential. In California, most varietal Durif wines are labeled Petite Sirah. Durif is often used as a blending component to add color and body to lighter red wines.

—Edward Weber

Gamay Noir

Synonyms

The many synonyms include Bourguignon noir, Petit Bourguignon, Gamay Beaujolais, and Petit Gamai in France, and Blauer Gamet in Germany. European Union legislation uses Gamay noir à jus blanc to avoid confusion among regions and reputed sources. There are a number of false Gamays found throughout the world. In California, they include Gamay Beaujolais, found to be a Pinot noir clone, and Napa Gamay, found to be Valdiguié.

Source

Gamay noir is now known to be a cross of Pinot noir and the ancient white variety Gouais, the latter a Central European variety that was probably introduced to northeastern France by the Romans. Gamay noir was grown in Burgundy for a long time, possibly as far back as the third century. It is an important variety of the Burgundy-Beaujolais region and Loire Valley in France and Valais, Switzerland. Otherwise, plantings of the true Gamay noir are very limited, including in California. The first true Gamay noir was imported into California in 1973.

Description

Clusters: small cylindrical; well-filled to compact; medium-size peduncles.
Berries: small to medium; short, oval shape; purple with bluish-white bloom.
Leaves: medium ; mostly entire with shallow lateral sinuses; V-shaped petiolar sinus; short, wide teeth; glabrous with few scattered hairs on lower leaf surface.
Shoot tips: downy, white tips; young leaves yellow with bronze-red highlights.

Growth and Soil Adaptability

The vine is moderately vigorous with a semi-upright growth habit. Budbreak is fairly early, but the fruitfulness of secondary buds reduces its vulnerability to spring frost. Its early ripening characteristics are best suited to cool climate regions. There are no reported limitations of soil types, but hillsides are often preferred to avoid highly productive, vigorous vines of lower fruit anthocyanins and tannins.

Rootstocks

There are no reported incompatibilities. In France, Riparia rootstocks are often used in granitic soils, 3309C in clay or clay-and-limestone soils, and 41B in extremely chalky soils.

clusters

Small cylindrical; well-filled to compact; medium-size peduncles.

berries

Small to medium; short, oval shape; purple with bluish-white bloom.

Clones

Earlier confusion about the true variety's presence and identity limited its importation and clonal diversity in California. Approximately 32 certified clones are registered in France, which is the origin of the material available in California. Gamay noir ENTAV-INRA® 358 is available as California certified stock. In addition, some generic French selections are available as California registered stock. According to data published in France, Gamay noir FPS 02 (French 221) is highly productive but of lower wine quality, Gamay noir FPS 03 (French 282) produces well-balanced wines, Gamay noir FPS 05 (French 509) produces wines of superior quality, and Gamay noir FPS 07 (French 284) vines are of high fertility. However, no replicated trial data has verified these results under California conditions. Most of these selections have received tissue culture therapy for virus infections that were present when the clones were imported, which may have altered the selections' performance.

Production

Gamay noir yields are moderate, averaging 4 to 7 tons per acre.

Harvest

Period: An early season variety, harvested in mid-September to mid-October in cool districts.

Method: Hand harvestability is medium-hard due to the short, woody peduncle. The medium-size clusters aid harvester productivity. There is no reported California experience with machine harvest of this variety.

leaves

Medium; mostly entire with shallow lateral sinuses; V-shaped petiolar sinus; short, wide teeth; glabrous with few scattered hairs on lower leaf surface.

Training and Pruning

Gamay noir is mostly pruned to bilateral cordons with 10 to 16 two-node spurs. Short pruning is universally recommended to avoid overcropping from the variety's high fertility of medium-size clusters.

Trellising and Canopy Management

Vertical-shoot-positioned systems are preferred in the cooler districts to improve fruit anthocyanin and phenolic content and, in conjunction with leaf removal, to minimize bunch rot.

Insect and Disease Problems

The variety is susceptible to fungal disease problems, including powdery mildew, Phomopsis cane and leaf spot, and Botrytis bunch rot. The compact clusters can contribute to bunch rot problems.

Other Cultural Characteristics

Gamay noir has characteristics similar to, and is most often grown in the same manner as, Pinot noir. It is recommended for medium-fertility soils. Shot berries from cool weather during bloom can be a problem. Excessive cluster exposure can cause sunburn. Vigorous vines produce a large second crop.

Winery Use

Because Gamay noir is traditionally used in light, fruity, uncomplicated red table wines for early consumption, a special fermentation process called carbonic maceration is common. Here the grapes are fermented whole with some pumping over to maximize pigments and minimize the tannins extracted from the skins. Beaujolais Nouveaux wines are largely made from this variety and vinification method. Red wines of more classical Burgundy style, as well as rosé wines, are also common to this variety.

—L. Peter Christensen

Gewürztraminer

Synonyms

Over 40 synonyms of Gewürztraminer exist, giving evidence of its worldwide popularity. Traminer, Rotclevner, Rousselet, Frenscher, and Edeltraube are used in France. It also has several synonyms in Germany, Austria, Switzerland, and Eastern Europe.

Source

Gewürztraminer is an aromatic variety grown throughout the world, most notably in Alsace, France. However, the variety may have originated in what is now Italy. Once grown in many parts of California, Gewürztraminer is now predominately planted in coastal counties in cool locations.

Description

Clusters: small; cylindrical with shoulders, often globular, compact; short peduncles.

Berries: small; short oval; unique tan-pink color and characteristic spicy "gewürz" taste.

Leaves: small; mostly entire with shallow lateral sinuses; closed V-shaped petiolar sinus with overlapping lobes; small teeth; bullate upper leaf surface; moderately dense, tufted hair on lower leaf surface.

Shoot tips: felty white with rose-pink margins; young leaves yellow-green with slight bronze highlights.

Gewürztraminer fruit is easily recognizable due to its pink to reddish color at maturity.

Growth and Soil Adaptability

Gewürztraminer has a trailing growth habit, and if planted on deep, fertile soils it may be vigorous depending on training and pruning. It is prone to poor fruit set, and thus is considered to be low yielding. Widely spaced rows should be avoided if possible in order to increase planting density and yield per acre. Gewürztraminer's early budbreak makes it sensitive to frost.

Rootstocks

Rootstock experience is limited due to the low acreage replanted to the variety in the late 1980s and 1990s. Gewürztraminer's vigor is site dependent, thus low- to moderate-vigor rootstocks should be considered in areas with deep, fertile soils. Soil conditions will help dictate the final rootstock choice.

Clones

Clonal evaluation has been conducted in Germany and France for many years with an eye toward selections with intense spicy characters. Three registered selections are available: Gewürztraminer FPS 01 and 02 are derived from an Alsatian selection (456), and FPS 03 came from a vine in California. Gewürztraminer ENTAV-INRA® 47 is available as California certified stock. No formal evaluations of these clones have been made in California.

clusters

Small; cylindrical with shoulders, often globular, compact; short peduncles.

berries

Small; short oval; unique tan-pink color and characteristic spicy "gewürz" taste.

Production

Gewürztraminer is not a highly productive variety since it is prone to coulure. Yields can vary considerably from year to year.

Harvest

Period: An early to midseason ripening variety, although winemakers often let the fruit hang for an extended period to develop more spicy characters.

Method: Gewürztraminer's short bunch stem makes hand harvesting difficult. Canopy shaking is easy to moderately easy. Fruit is removed mostly as single berries and some cluster parts, with juicing light to medium. Canopy shaking can cause moderate shoot breakage. Trunk shaking is easy to moderately easy, with less MOG. Fruit is removed as single berries, but with more cluster parts than with canopy shaking. Juicing is light.

Training and Pruning

Due to its small clusters and high vigor, Gewürztraminer should be head trained and cane pruned in order to leave an adequate number of buds. However, many Gewürztraminer vineyards in California are cordon trained and spur pruned. When cordon trained and spur pruned, shoot thinning may be minimal, making mechanical pre-pruning advisable.

Trellising and Canopy Management

For low- to moderate-vigor sites, vertical-shoot-positioned systems are appropriate. On high-vigor sites, split canopy systems can be used to increase the yield potential and balance vegetative growth. Because fruit set is variable, shoot thinning is often delayed until after set or is not performed at all so that all shoots with clusters are kept in order to maximize yield. Leaf removal can be used to reduce the risk of Botrytis bunch rot.

leaves

Small; mostly entire with shallow lateral sinuses; closed V-shaped petiolar sinus with overlapping lobes; small teeth; bullate upper leaf surface; moderately dense, tufted hair on lower leaf surface.

*Felty white with rose-pink margins;
young leaves yellow-green with
slight bronze highlights.*

Insect and Disease Problems

Gewürztraminer's small, tight clusters make it
susceptible to Botrytis bunch rot. Older vineyards commonly carry virus disease.

Other Cultural Characteristics

Gewürztraminer's early leafing habit makes it vul-
nerable to spring frosts. Varietal character develops late in the ripening period. Harvesting too
early results in wines lacking in varietal character. Gewürztraminer should only be grown in
cool regions due to its low acid content.

Winery Use

Gewürztraminer produces distinctive wines with
a spicy, floral aroma. Table wines are usually
slightly sweet to offset its natural bitterness.
Excellent dessert wines can also be produced
from this variety.

—*Rhonda J. Smith and Edward Weber*

Grenache

Synonyms

In Spain the cultivar is known as Garnacha, and in France it is called Grenache noir. It is also referred to as Cannonao or Cannonaddu on the island of Sardinia, and in southern Italy as Granaccia or Alicante.

Source

Grenache is one of the most widely planted wine grape varieties in the world, with vast acreages encompassing the southern Mediterranean wine region. It is generally believed to have originated from the northern Spanish province of Aragon. The variety spread from this area to the Rioja and Navarre regions, then moved both north and south of the Pyrenees Mountains. Grenache was first planted in France in the Longuedoc region in the early eighteenth century, and it reached the southern Rhône Valley by the nineteenth century. In the southern Rhône, Grenache is generally blended with Syrah and other varieties to produce common red table wines (Cotes-du-Rhône) as well as the highly regarded Chateauneuf-du-Pape.

Grenache was probably first introduced to California by Charles Lefranc, a prominent Santa Clara wine grower, in 1857. Its versatility made it popular in the planting boom in the late 1800s. Acreage grew steadily after Prohibition, especially in the Central Valley for dessert and rosé table wines. It is now grown in diverse climatic regions for blending and the production of varietal red and blush table wines.

Description

Clusters: medium to large; broad conical, well-filled to compact; medium-length peduncles.

Berries: small to medium; round to short oval; purple, much lighter in sunlight.

Leaves: medium; mostly entire with shallow superior lateral sinus; narrow U-shaped petiolar sinus; short, sharp teeth; upper leaf surface very smooth, waxy (like wax paper) green; glabrous on lower surface.

Shoot tips: downy tips; young leaves mostly green with slight bronze-red highlights.

Growth and Soil Adaptability

The vine has potentially high vigor in medium- to fine-textured soils (sandy loam to clay loam) and low vigor on sandy soils. It is more vigorous than Chardonnay, but generally less vigorous than Cabernet Sauvignon. Canes are thick with an upright growth habit. Own-rooted plantings on shallow soils should be avoided in the San Joaquin Valley.

Rootstocks

No known incompatibilities exist with commonly used rootstocks. In coastal regions, where phylloxera resistance is desired, Grenache is generally grafted onto moderate-vigor rootstocks such as 101-14 Mgt and 3309C. In hillside plantings, or in areas where soil depth or fertility is limited, 110R and 1103P may be used. In the San Joaquin Valley, where nematode resistance is desired, Freedom and Harmony are commonly used.

clusters

Medium to large; broad conical, well-filled to compact; medium-length peduncles.

berries

Small to medium; round to short oval; purple, much lighter in sunlight.

Clones

There are two old California registered selections available of Grenache (now known at FPS as Grenache noir to avoid confusion with other forms of Grenache). Recent studies indicate that Grenache noir FPS 01A, a field selection from California, is fruitful, has smaller berries, and exhibits less propensity for bunch rot than FPS 03, a field selection from the UC Jackson Foothill Experiment Station. Although selection 03 produces larger yields than selection 01A, because its berries and clusters are larger and more prone to rot and its fruit maturation is delayed, it is not recommended. The diversity of Grenache planting stock in California has increased dramatically with new imports from Italy and France. Grenache noir FPS 04 (Rauscedo VCR 3) is available as California certified stock. Grenache noir ENTAV-INRA® 70, 136, 362, 513, and 515 are also commercially available in California.

Additional testing is needed, particularly in coastal regions, to determine the potential merits of these selections.

Production

Grenache is a consistent producer, with yields that may range from 4 to 8 tons per acre in coastal regions and 8 to 14 tons per acre in the San Joaquin Valley.

Harvest

Period: A midseason ripening variety, typically harvested from mid- to late September in the San Joaquin Valley. In some cases harvest may be earlier if a lower alcohol rosé or blush wine is produced. In coastal regions, it is harvested from mid-September to mid-October.

Method: Clusters have thick peduncles, requiring that knives or shears be used for fruit removal with hand harvest. Grenache is difficult to machine harvest with canopy shakers, with most fruit removed as single berries and with medium to heavy juicing. Its large vine framework interferes with rod penetration, and dead spur removal is a problem. Harvest with trunk shakers is moderately difficult, with medium juicing. Most fruit is removed as single berries. Considerable force is required to remove the fruit.

leaves

Medium; mostly entire with shallow superior lateral sinus; narrow U-shaped petiolar sinus; short, sharp teeth; upper leaf surface very smooth, waxy (like wax paper) green; glabrous on lower surface.

shoot tips

Downy tips; young leaves mostly green with slight bronze-red highlights.

Training and Pruning

Grenache is commonly trained to bilateral cordons and pruned to 12 to 16 two- to three-node spurs per vine. Many older vineyards are head trained and spur pruned (14 to 18 two- to three-node spurs retained per vine). Quadrilateral cordon training and spur pruning (approximately 24 two- to three-node spurs retained per vine) is used to a limited extent in the San Joaquin Valley where anticipated vine vigor is high due to deep, fertile soils and vigorous rootstocks. If quadrilateral cordon training is used, careful attention must be paid to avoid overcropping.

Trellising and Canopy Management

Clusters are subject to bunch rot at harvest, thus basal leaves may be removed following fruit set to improve the fruit zone microclimate. In coastal regions shoot thinning is also performed to reduce canopy density and decrease crop load. Its upright growth habit makes this variety well suited for vertical-shoot-positioned systems in coastal regions. The traditional California two-wire vertical trellis is commonly used in the northern and central San Joaquin Valley.

Insect and Disease Problems

Grenache is highly susceptible to Eutypa dieback and moderately susceptible to Phomopsis and Botrytis shoot blight diseases in cool, wet springs. Its moderately compact clusters are prone to bunch rot near harvest.

Other Cultural Characteristics

Vine stress due to overcropping, insect damage, or other factors may result in delayed, erratic budbreak the following spring. Overcropping may also cause the vines to enter an alternate-bearing pattern in which yields fluctuate drastically from year to year. Wood maturity may also be a problem in highly vigorous vineyards, particularly if vines grow late into the season. Winter injury can be a problem with vigorous or overcropped vines during the second through fourth years of vine training. Poor light exposure into the fruit zone during bloom, caused by dense canopy growth, may reduce fruit set and result in excessively loose or straggly clusters.

Winery Use

Wine color, body, and aging potential are typically low when fruit is grown in warm regions. Grenache is used primarily for the production of varietal rosé and blush wines in the San Joaquin Valley. It is commonly blended with Syrah, Mourvèdre, or other varieties for the production of Rhône-style red table wines in coastal regions.

—*Nick K. Dokoozlian*

Malbec

Synonyms

Malbec's official name is Cot in France, where it has many synonyms, including Auxerrois, Malbeck, Noir de Pressac, Cote rouge, Cahors, Cot de Pays, Jacobain, and Pied Rouge. It is known in southwest France and the Loire Valley as Cot. In the New World, including Argentina, Chile, Australia, New Zealand, and the United States, it is known as Malbec.

Source

The variety's origin is uncertain. Since the eighteenth century it has been known in southwestern France, where it is still mainly grown in that country. It is a minor variety in Bordeaux and the Loire Valley. It is also grown in Italy, Argentina (where it is the third major variety), Chile, Australia, New Zealand, and California. In 1858, Charles Lefranc brought it from France to the Santa Clara Valley, as did Jean-Baptiste Portal in 1872. California acreage declined sharply in the late 1800s due to phylloxera. It was planted little thereafter until recently. Now there is renewed interest in the variety for blends of Cabernet Sauvignon and Merlot and for varietal wines.

Description

Clusters: medium; wide conical, loose to slightly compact; short to medium peduncles.

Berries: medium; round; purple-black.

Leaves: medium; mostly entire to slightly 3-lobed with reduced lateral sinuses; U-shaped petiolar sinus; short, sharp teeth; lower leaf surface is glabrous to covered with sparse, tufted hair.

Shoot tips: felty tips; young leaves felty white to downy with bronze-red highlights.

Growth and Soil Adaptability

Malbec is a vigorous variety adaptable to a wide range of soil types. It is sensitive to coulure or shelling, especially with high vigor or cool weather during bloom.

Rootstocks

Rootstock selection should be based on the type of soil pests present, the potential vigor of the site, vine spacing, and desired vine size. High-vigor rootstocks such as St. George, Freedom, or 1103P should not be used due to Malbec's tendency to coulure. Moderate- to low-vigor rootstocks such as 3309C, 101-14 Mgt, and SO4 might be reasonable choices.

clusters

Medium; wide conical, loose to slightly compact; short to medium peduncles.

berries

Medium; round; purple-black.

Clones

Clonal selection is important in this variety to minimize the risk of coulure. Malbec FPS selections 04 and 06 are particularly prone to poor fruit set and low yields. Malbec FPS 08 (non-registered) is consistently higher yielding but is still moderate in production. Recently made available, Malbec FPS 09 (French Cot 180) and 10 (French Cot 46) have potential for more consistent crop set.

Production

Poor fruit set is a problem in Malbec, and some vineyards may yield only 1 to 3 tons per acre. If set is good, production levels are moderate to high.

Harvest

Period: A late-midseason variety.
Method: Canopy shaking is easy, with juicing medium to heavy.

Training and Pruning

The variety is usually cordon trained and spur pruned. Low-yielding clones may benefit from cane pruning in order to increase the cluster number. Attention should be paid to adequate vine spacing to balance Malbec's vegetative growth tendency.

leaves

Medium; mostly entire to slightly 3-lobed with reduced lateral sinuses; U-shaped petiolar sinus; short, sharp teeth; lower leaf surface is glabrous to covered with sparse, tufted hair.

shoot tips

Felty tips; young leaves felty white to
downy with bronze-red highlights.

Trellising and Canopy Management

Malbec's strong, upright growth makes it particularly suited to vertical-shoot-positioned systems, particularly for low- to moderate-vigor sites. Malbec has large leaves and unusually strong lateral shoot growth, leading to a dense canopy in the fruit zone. Leaf and lateral shoot removal are recommended to improve fruit quality and to reduce the risk of Botrytis bunch rot. On higher-vigor sites, split canopy systems can be used to increase the yield potential and balance the vegetative growth.

Insect and Disease Problems

Malbec is moderately sensitive to powdery mildew. Its high vigor may encourage grape leafhoppers.

Other Cultural Characteristics

Certain clones are especially prone to coulure. In some years this can lead to particularly low yields.

Winery Use

Malbec is usually used in small amounts to blend into Cabernet Sauvignon and Merlot wines. As a varietal, it makes a softer red wine with limited aging potential but very high quality.

—*Edward Weber*

Malvasia Bianca

Synonyms

In France, the variety is known as Malvasia and Malvoisie; in Germany, it is called Malvasier and Früher Roter Malvasier; and in Italy it is referred to as Malvasia bianca del Chianti, Malvasia di Candia, Malvasia Rosso, Malvasia del Lazio, Malvasia Puntinta, and Uva Greca. Spanish synonyms include Malvasia Fina, Rojal, Subirat, Blanquirroja, Blancarroga, Tobia, Cagazal, and Blanca-Roja. It is known as Malvasia Fina in Portugal, Malvasia Candid in Madeira, Malvazija in the former Yugoslavia, and Monemvasia in Greece.

Source

Currently Malvasia bianca is understood to be from the northwest coast of Italy where it is an obscure variety there known as Malvasia bianca piemonte or Moscato greco. Its origin within Italy and beyond is unknown.

Description

Clusters: medium; long conical and shouldered, well-filled to compact; medium peduncles.

Berries: medium; round; yellow to oily brown when ripe; muscat flavor.

Leaves: medium; deeply 5-lobed with lyre-shaped to overlapping petiolar sinus; large, sharp, jagged teeth; lower surface with moderately dense hair.

Shoot tips: cobwebby hairs on tip; green, young leaves with bronze-red highlights, glabrous and shiny.

Growth and Soil Adaptability

Vines have moderately high vigor and are moderately productive to highly productive. They grow best on well-drained soils of moderate texture to somewhat coarse-textured sandy soils. On very coarse sand, nutrient deficiencies can occur and berry set can be negatively affected. In southern San Joaquin Valley berry set can be a problem on all soil types, and may be worse when the vines are vigorous. Sandy soils with low zinc availability may also increase berry set problems. Magnesium deficiency symptoms can easily occur on sandy soils or sites prone to excessively wet spring conditions. Vine spacing should be about 7 to 9 feet or more for vertical-shoot positioning or standard bilateral cordons. For horizontally divided quadrilateral vines, spacing should be 6 to 7 feet down the row.

Rootstocks

Moderate vigor rootstocks are probably best, such as 101-14 Mgt, Kober 5BB, or 1103P. Vigorous rootstocks may increase susceptibility to poor berry set and reduce yields. Freedom and Ramsey have been common in the past, but should be used with caution.

clusters

Medium; long conical and shouldered, well-filled to compact; medium peduncles.

berries

Medium; round; yellow to oily brown when ripe; muscat flavor.

Clones

There are limited registered clones available. Currently there is only one FPS-registered selection, Malvasia bianca FPS 03. The importation of new clones from Italy is desirable, as FPS 03 is of questionable trueness to variety, according to Italian ampelographers.

Production

Malvasia bianca's vines are moderately to very productive, capable of bearing large crops—8 to 12 tons per acre. Higher yields may sacrifice some flavor intensity and delay harvest. Yields of Malvasia bianca can be variable and disappointing due to poor set and low bud fruitfulness. The variety can sunburn easily if it is overexposed, and its fruit tends to amber easily with high maturity levels.

Harvest

Period: An early to midseason variety, ripening in late August to late September.

Method: Large clusters, easily cut peduncle, and productivity make hand harvest easy. The vine is somewhat adapted to trunk shaker type heads, but machine harvest is still difficult, with medium potential for juicing of berries at harvest. Machine harvest by pivotal striker is less desirable, while newer, bow-rod heads may reduce the possibility of juice losses.

leaves

Medium; deeply 5-lobed with lyre-shaped to overlapping petiolar sinus; large, sharp, jagged teeth; lower surface with moderately dense hair.

shoot tips

Cobwebby hairs on tip; green, young leaves with bronze-red highlights, glabrous and shiny.

Training and Pruning

Malvasia bianca is well suited to spur pruning and bilateral cordon or quadrilateral training. A spur count of 14 to 18 two-node spurs is acceptable, depending on rootstock, soil depth, and soil texture. Cane pruning is not suggested without cluster thinning or severe overcropping may result with reduced shoot vigor and delayed harvest.

Trellising and Canopy Management

A standard bilateral cordon with a cross-arm foliage wire is recommended. Vertical-shoot-positioned systems are also acceptable. High-vigor sites may benefit from a GDC-type trellis but cross-arm foliage wire may again be required to avoid excessive ambering of fruit.

Insect and Disease Problems

Bunch rot can be a problem due to the relatively thin berry skin, but severe rot is not common with moderate irrigation levels and nitrogen application. It is somewhat resistant to powdery mildew. Leafhoppers can be a problem in some cases, especially as an annoyance to hand-pickers at harvest.

Other Cultural Characteristics

At optimum maturity amber, slight browning, and some dark spotting may occur as with many muscat types. Coarse soils and severe land leveling may induce magnesium deficiency symptoms in leaves. Fill areas tend to show symptoms as opposed to cut areas seen with potassium problems. At moderate- to high-crop loads, sugar levels may not exceed 19 to 20° Brix at full maturity, with less intensity of muscat flavors.

Winery Use

Highly flavored wines of good to excellent quality and moderate acidity are produced. These wines are used as a main base or as a blend in sparkling wines, as an enhancement of fruit character in Chardonnay wines, or for dessert-type premium wines. Future interest as a sparking wine base or premium dessert wine may increase but still may be somewhat limited to niche markets and winery specialty needs.

—Paul S. Verdegaal

Melon

Synonyms

In France, the variety is known as Melon de Bourgogne and Muscadet and in Germany as Weisser Burgunder or Später Weisser Burgunder. Melon de Bourgogne is an approved synonym in the United States. In California it was misidentified as Pinot blanc.

Source

Melon is an old Burgundian variety where it was once widely grown. In France, Melon is mainly used for the production of Muscadet wines in the Loire Valley. In California, it is a minor variety with small plantings in the cooler production areas of the North and Central Coast regions. Prior to the introduction of true-to-type selections in the 1980s, most Pinot blanc vineyards in California were plantings of Melon.

Description

Clusters: small to medium; compact, conical to cylindrical; short peduncles.

Berries: small to medium; round; yellow with white bloom; prominent lenticels; skins with high tannin.

Leaves: medium; mostly entire with relatively closed U-shaped petiolar sinus; upper surface bullate; short, rounded teeth; lower surface mostly glabrous with very sparse, tufted hair.

Its leaves are similar to those of Chardonnay and distinguished by not having "naked" veins along petiolar sinus.

Shoot tips: felty cream-yellow; young leaves yellow-green. Pinot blanc shoot tips and young leaves are felty white, a characteristic that can distinguish it from Melon in the spring.

Growth and Soil Adaptability

Vine vegetative growth can vary significantly from weak to moderately vigorous depending on the climatic region, soil characteristics, moisture availability, and rootstock selection. Adaptable to a wide range of soils, Melon's highest vigor will be on deep loam or clay loam soils with high moisture availability. Shoot growth on non-positioned canopies is semi-erect. Melon has early budbreak, which makes it sensitive to frost.

Rootstocks

Rootstock experience is limited due to the low acreage replanted in the 1980s and 1990s. Melon has no known incompatibilities when certified budwood is used to propagate the planting stock. Rootstock selection should be based on soil characteristics, the pests present, the potential vigor of the site, vine spacing, and desired vine size. Rootstocks may have more influence on sites where anticipated vigor is low, and the choice may have a greater effect on vine growth and development.

clusters

Small to medium; compact, conical to cylindrical; short peduncles.

berries

Small to medium; round; yellow with white bloom; prominent lenticels; skins with high tannin.

Clones

There are five registered selections of Melon. FPS 01 was selected from Beaulieu Vineyards. Melon FPS 05 came from Inglenook's Napa Valley Vineyards; selection 07 is a heat-treated subclone of 05. In France, there are 10 certified ENTAV clones. One of them, Melon ENTAV-INRA® 229, is now available in California and is a registered selection at FPS.

Production

Vine yield can vary considerably by climatic region, site influences, bunch rot level, and cultural practices. Melon is a moderately productive variety; crop size can range from 3 to 6 tons per acre.

Harvest

Period: A midseason variety, harvested from mid-September to early October in the coastal regions.

Method: Hand harvest with knives or shears is easy due to the lack of excessive canopy growth. Horizontal rod or bow machine harvest is medium in difficulty with fruit coming off mostly as single berries with moderate juicing. Bow-rod picking heads used on well-trained vines on vertical-shoot-positioned trellises have lower shoot and spur breakage than straight rods. Trunk shaker machine harvest is medium in difficulty with fruit coming off as single berries with medium juicing, although brittle wood can be a problem.

Training and Pruning

Melon is commonly trained to bilateral cordons and spur pruned. In very cool regions where bud fruitfulness is low, head training and cane pruning may result in higher productivity. For low-vigor sites, higher plant densities and the use of unilateral cordon training may produce vines that better balance fruit and vegetative growth.

leaves

Medium; mostly entire with relatively closed U-shaped petiolar sinus; upper surface bullate; short, rounded teeth; lower surface mostly glabrous with very sparse, tufted hair.

shoot tips

Felty cream-yellow; young leaves yellow-green. Pinot blanc shoot tips and young leaves are felty white, a characteristic that can distinguish it from Melon in the spring.

Trellising and Canopy Management

For low- to moderate-vigor sites, vertical-shoot-positioned systems are appropriate. The use of split canopy systems should be considered only on sites with especially high potential vigor. Leaf removal in the fruit zone can be used to reduce the risk of Botrytis bunch rot.

Insect and Disease Problems

The compact clusters are susceptible to bunch rot, mainly by *Botrytis cinerea*. Orange tortrix, *Argyrotaenia citrana*, prefers compact-clustered varieties, and high populations have been observed in some Melon vineyards.

Other Cultural Characteristics

Vine growth can be slow and irregular, and the leaves can show a temporary chlorosis when early spring temperatures are cold. Crop recovery after spring frost is good.

Winery Use

Melon can produce table wines of good flavor and balance when grown in the cooler coastal regions. The grapes have also been used to make base wine for sparkling wine production. Due to the higher tannin content in the skins, the wines are more prone to browning if not handled correctly.

—*Larry J. Bettiga and Edward Weber*

Merlot

Synonyms

Merlot is the common name outside of France, while Merlot noir is officially used in France. French regional synonyms include Merlau rouge, Crabutet noir, or Plant Medoc (Bazadais); Alicante (Podensac); Seme dou flube (Graves); Seme de la Canau (Portes); Semilhoun rouge (Medoc); and Bordeleze belcha (Basque country); and it is Medoc noir in Hungary.

Source

Little is known of the origin of the variety, but it has been cultivated in the Bordeaux region since the eighteenth century. The first true botanical description was in 1854 by V. Rendue who described it favorably for blending with Malbec and Cabernet Sauvignon and as a component of the great wines of Medoc. A resurgence of planting in France since the 1970s, particularly in the south, makes it the third most planted black variety there. Antoine Delmas imported the first vines to California in the 1850s; only a few acres existed after Repeal. It was included in the California planting boom of the 1970s, and plantings soared after 1987. Merlot acreage grew faster than that of any other world-class variety in the 10 years that followed with the exception of Viognier. It is also widely planted in Italy, Central Europe, and South America.

Description

Clusters: small to medium; long cylindrical with large shoulders, well-filled; short to medium peduncle.

Berries: small; round; blue-black with whitish bloom; green rachis prominent.

Leaves: medium (often very large in training years); deeply 5-lobed, longer than wide; lateral sinuses often overlapping and occasionally with teeth at base; wide U-shaped petiolar sinus; narrow, sharp teeth; slight tufted hair on underside of leaves.

Shoot tips: felty white with rose margin; young leaves cream-yellow and downy.

Merlot is distinguished from Cabernet Sauvignon by young leaf color (cream-yellow versus bronze red), petiolar sinus, and cluster shape.

Growth and Soil Adaptability

Merlot has medium-high vigor with a trailing growth habit. Excess vigor quickly creates a dense canopy due to lateral shoot development. It is adapted to cool to warm climate regions. Merlot does well on deep, sandy loam or well-drained soils that have good moisture-holding capacity.

Rootstocks

There are no known incompatibilities; rootstock selection can be largely based on site and soil conditions and cultural requirements. High-vigor rootstocks such as Ramsey and O39-16 should be avoided unless nematode or fanleaf virus resistance is imperative. Freedom should be used with caution but is acceptable for quality production where nematodes are a severe problem. These rootstocks, as well as St. George, may influence high tissue nitrogen levels in the scion that may contribute to fruit set problems. When

clusters

Small to medium; long cylindrical with large shoulders, well-filled; short to medium peduncle.

berries

Small; round; blue-black with whitish bloom; green rachis prominent.

compared to own-rooted vines, any rootstock reduces the tendency for high nitrogen levels in the vine. High nitrogen is thought to contribute to coulure and early bunch stem necrosis.

Clones

Merlot FPS 03 (an Inglenook Vineyard selection with heat treatment) is a California standard due to its consistency of fruit set, yield, and fruit composition. Selections FPS 01 (Inglenook, no heat treatment) and FPS 06 (Monte Rosso, heat treated) have performed similarly to selection 03. Selection FPS 08, an introduction from Argentina, is of lower yield and fruit-to-pruning weight ratio due to poorer fruit set, especially with cool weather. Clonal introductions from Italy and France have added to clonal diversity and are under evaluation. So far, FPS selection 9 (Rauscedo 3) appears to be very similar to selections 1 through 8. Little information is available

on the California performance of Merlot FPS 10, 11, 12, 13, and 21 (all from Conegliano, Italy); Merlot FPS 14 (French 348), 15 (French 181), 19 (French 343), and 25 (French 314). Merlot FPS 18, the Bear Flats clone developed by Sterling Winery and donated to the public University of California collection, is a popular heritage selection. Merlot ENTAV-INRA® 181, 346, and 348 are registered selections at FPS. In addition, ENTAV-INRA® 182, 314, 343, and 347 are available commercially in California through the ENTAV-INRA® trademark program.

Production

Yields tend to vary from year to year. Production is good to very good: 3 to 7 tons per acre in coastal regions and 5 to 9 tons per acre in the interior valleys.

Harvest

Period: A midseason variety, ripening in mid-September to mid-October.

Method: Hand harvest is easy as peduncles are long and easily cut. Canopy shaking results in medium harvestability, but some reports range from easy to hard and light to medium juicing. Most fruit (about 70%) is removed as single berries with some whole clusters and parts. Trunk shaking results in medium harvestability, but with a range of easy to medium-hard and light juicing. Fruit is mostly

leaves

Medium (often very large in training years); deeply 5-lobed, longer than wide; lateral sinuses often overlapping and occasionally with teeth at base; wide U-shaped petiolar sinus; narrow, sharp teeth; slight tufted hair on underside of leaves.

shoot tips

Felty white with rose margin; young leaves cream-yellow and downy.

removed as single berries with some clusters. The addition of straight rods may be needed if clusters are not full. With all harvester types, the fragile, brittle shoots tend to break, and flaccid, ripe berries can be a problem with removal.

Training and Pruning

Vines are most commonly trained to bilateral cordons and pruned to 14 to 20 spurs with two to three nodes each. Quadrilateral cordon training can be used in medium- to high-vigor sites but with care to avoid overcropping.

Trellising and Canopy Management

Merlot is usually not quite as vigorous as Cabernet Sauvignon but has similar canopy management requirements. Vertical foliage support is used in low- to medium-vigor vines, while Smart-Dyson, lyre, and GDC systems are suited to medium-high– to high-vigor vines.

Insect and Disease Problems

Vineyards planted in the 1970s often experienced erratic fruit set and low yields due to poor or virus-diseased wood sources and the use of own-rooted vines. Merlot is slightly susceptible to powdery mildew and Pierce's disease and is moderately susceptible to Botrytis bunch rot with fall rains. It is somewhat resistant to Eutypa dieback.

Other Cultural Characteristics

Merlot is susceptible to poor fruit set if cool weather occurs during bloom, which often contributes to seasonal variations in productivity. Its own-rooted vines tend to accumulate high levels of nitrogen compounds, including nitrates, during bloom, especially during cool weather. Thus, judicious and moderate nitrogen fertilization is recommended; post-bloom applications are advisable. The use of resistant rootstocks tends to minimize or even eliminate this problem. Merlot is somewhat sensitive to soil problems that involve zinc deficiency, salinity, and cold, excessively wet conditions.

Winery Use

Historically, Merlot was primarily used for blending with Cabernet Sauvignon and other Bordeaux varieties to add softness and fruit complexity, shorten aging requirements, and to hedge the risk of cool, late-ripening conditions in Bordeaux. In recent years it has also become popular as a full-bodied, high-quality varietal wine that can be marketed sooner than Cabernet Sauvignon.

—*L. Peter Christensen*

Mourvèdre

Synonyms

In France, the names Mourvède, Mourvedon, Mourves, Morvede, Morvegue, and Mourveze that are used in Provence relate to the color of the grape, as do the names Negron, Negre Trinchera, or Trinchiera in the Drôme. It is also known as estrangle-chien ("dog-strangler"), due to the hard, astringent taste of its fruit; Buona Vise ("good cane"); and Tire Droit, for the vertical direction of its canes, in the Drôme. It is called Espar or Spar in Herault; Planta de Saint-Guilles in Gard; Catalan in Bouches-du-Rhône; Baltazar in Gironde; Benadu, Negron, and Piemonaise in Vaucluse; and Flouron or Flouroux in Ardèchen, indicating the heavy bloom on its grapes. In Spain, it is called Monastrell, Tinto, Tinta, Tintilla, Alicante, and Mataró. In the United States, it has been widely known as Mataro.

Source

Wine historians suspect that the variety is of ancient origin, perhaps introduced to the Barcelona area of Spain by the Phoenicians in 500 BC. The name Mourvèdre is derived from the town of Murviedro in Valencia, and the name of Mataro is derived from the town of Mataró in Catalonia. After the sixteenth century, the variety was brought to France. The grape is thought to have arrived in California in the 1860s in the Pellier collection, a consignment of stock from France to Santa Clara Valley by Louis and Pierre Pellier. It was popular in the Santa Clara Valley in the 1870s, and by the end of the century it was included in Zinfandel vineyards in the North Coast as part of a field blend. Vineyards were also planted in Contra Costa about that time, and several are still in production. It was also popular in Riverside and San Bernardino until the urbanization of those areas beginning in the 1950s. Spain is dominant in planted acreage, with lesser plantings in Algeria, France, Australia, and California. There are also small amounts in Tunisia and the former Soviet Union.

Description

Clusters: medium; broad conical, often winged; well-filled to compact; short to medium peduncle.

Berries: medium; round, blue-black with distinct white bloom; juicy pulp with a harsh taste.

Leaves: medium; mostly entire with U-shaped petiolar sinus and very shallow superior lateral sinuses; short, sharp teeth; dense hair on lower surface.

Shoot tips: open, felty tips; young leaves are yellow-green with slight bronze highlights.

Growth and Soil Adaptability

In Spain and France, it is planted in deep, well-drained soils that ensure a regular supply of moisture; without irrigation, it is prone to drought stress. Most of the plantings are within 50 miles of the Mediterranean Sea, which are areas without severe winters. In California, the best wine quality has been achieved on deep, sandy soils of low fertility and minimal irrigation. Mourvèdre requires

clusters

Medium; broad conical, often winged; well-filled to compact; short to medium peduncle.

berries

Medium; round, blue-black with distinct white bloom; juicy pulp with a harsh taste.

considerable heat to adequately ripen fruit, especially in the period between veraison and harvest. This variety is probably best adapted to a high Winkler Region III or low Winkler Region IV. It is planted principally in Contra Costa County near the confluence of the Sacramento River with San Francisco Bay.

Like many varieties of Spanish origin, Mourvèdre is upright and vigorous in its growth. In California, older plantings are head pruned, and spaced 8 by 8 feet. Newer plantings in the North and Central Coast regions are spaced 6 by 8 feet. In the spring, its shoots can be broken by strong winds. Fruit clusters are medium to large, and there is good basal bud fertility, which makes this variety well suited to spur pruning. In districts with cold winters (such as Lake County), winter injury can be severe. In the foothills this variety can be prone to sunburn.

Rootstocks

In Spain and France, 110R and 41B are used on soils of moderate to high lime content. In California, some vineyards are planted on their own roots, along with AXR #1, St. George, and others. The rootstock incompatibilities reported in French literature have been alleviated with better wood sources, suggesting virus problems in earlier wood sources.

Clones

Originally, 15 clones were selected between 1973 and 1976 from southern France by ENTAV and INRA. Mourvèdre ENTAV-INRA® 233 and 369 are presently available commercially in California. As "Mataro," FPS offers registered selection Mataro 01, which was selected from California vineyards, and FPS 03, which was derived from FPS 01 using heat therapy.

Production

The majority of Mourvèdre vineyards in production in California are old, head-pruned vines that are minimally irrigated. Statewide average production is 3 tons per acre. Newer vineyards planted in higher densities with cordon training and vertical-shoot-positioned trellises appear to be more productive, with yields of 5 to 6 tons per acre in the North Coast.

leaves

Medium; mostly entire with U-shaped petiolar sinus and very shallow superior lateral sinuses; short, sharp teeth; dense hair on lower surface.

shoot tips

*Open, felty tips; young leaves are
yellow-green with slight bronze
highlights.*

Harvest

Period: Late, harvested in mid- to late October.
Most seasons, Mourvèdre is among the last
varieties to be harvested. It has an unusual
characteristic of the fruit becoming dimpled
like a golf ball as it approaches harvest.

Method: The upright shoots and medium to
large clusters make this an easy cultivar to
hand harvest. Machine harvesting is not commonly done in the United States. In Europe, it
is successfully machine harvested by canopy
shakers using bow rods. The berries separate
readily from the clusters.

Training and Pruning

Older vineyards are head trained with spur
pruning. Shoot thinning is widely practiced to
remove nonproductive shoots, increase air flow
to discourage bunch rot and powdery mildew,
and to improve fruit color. Newer vineyards are
cordon trained on vertical-shoot-positioned
trellis systems.

Trellising and Canopy Management

This cultivar is well-suited to vertical-shoot-
positioned systems, due to its upright growth.
Cordon wires are set between 30 and 40 inches.
Two sets of movable foliage wires are used to
keep shoots upright. In Europe, VSP systems are
widely used, but the vines are smaller in stature
and more closely spaced.

Insect and Disease Problems

Mourvèdre is sensitive to mites, grape leafhoppers, and esca (in Europe). The thick skin
helps to resist Botrytis bunch rot. Tight clusters
can contribute to sour rot in warm districts.

Other Cultural Characteristics

This cultivar breaks bud late and ripens late.
Consequently, it needs to be planted in warm
sites to adequately ripen.

Winery Use

Mourvèdre is used to make both fruity rosés and
concentrated, dark-red wines with strong tannic
structure. Due to a strong concentration of
antioxidants, the wines age well, and they are
often used for blending with wines more prone
to oxidation, such as Grenache in the southern
French appelation of Chateauneuf-du-Pape.
These wines benefit from oak aging, especially if
the yield per vine is limited.

—Glenn McGourty

Muscat Blanc

Synonyms

The official French name is Muscat à petits grains, which means simply "Muscat with small berries." In California, the name Muscat Canelli is common, which is a modification of Moscato di Canelli, a geographical derivation used in Italy. Other names that reference a district where the grapes are grown include Moscato d'Asti in Italy and Muscat de Frontignan in France. Muscat blanc in California describes the white-fruited selection that is almost exclusively grown here.

Source

Likely a native of Greece, Muscat blanc has been cultivated on the edge of the Mediterranean Sea since ancient times. The Romans probably brought the first vines to Narbonne, France, where they became a notable variety in places such as Frontignan. It remains an important variety in Italy for sparkling and dessert wines, and it is widely grown throughout Europe and the New World. The three color variants of the variety—white, rosé, and red—are the result of mutations of berry skin color. The white form predominates in Europe as well as California. It was brought to California in the 1850s from nurseries in New England, where it was grown as a hothouse table grape. It is now grown in widely differing districts in California, owing to its use as dessert wine in warm districts and light, sweet table wines in the cooler districts.

Description

Clusters: medium; cylindrical to conical, well-filled to compact; medium peduncles.
Berries: medium; round; yellow and oily brown at maturity.
Leaves: medium; mostly 3-lobed to almost entire with reduced lateral sinuses; closed U-shaped petiolar sinus; sharp, saw-like teeth; lower leaf surface glabrous.
Shoot tips: downy tips; young leaves bronze-red and shiny.

Growth and Soil Adaptability

The vine is moderately vigorous on medium- to fine-textured soils—sandy loams to clay loams—when grown on its own roots in the San Joaquin Valley. Vigor is poor on sandy soils. In-row vine spacing is 6 or 7 feet, and row width spacing can be 8 to 10 feet for conventional equipment.

Rootstocks

Freedom, Harmony, and Ramsey are suitable rootstocks in nematode-prone sites. Phylloxera-resistant rootstocks should be of medium to high vigor to produce an adequate canopy. Otherwise, vine growth may be insufficient, resulting in an excessively open, sparse canopy.

clusters

Medium; cylindrical to conical, well-filled to compact; medium peduncles.

berries

Medium; round; yellow and oily brown at maturity.

Clones

Muscat blanc FPS 01 and 02 (both 64-day heat treatments), are the most widely distributed selections in California, along with other commercial wood sources. Clonal studies show that selections FPS 03 and 04, introductions from Milan, produce more and smaller clusters with fewer and smaller berries and less rot. The Milan selections also demonstrate the largest harvest yields of comparable fruit composition. Of these four tested selections, 04 is recommended for new plantings due to its greatest cluster numbers and much less bunch rot (50 percent less in a trial at the UC Kearney Agricultural Center than selection 02). Selection 04 fruit maturation was also earlier than that of selection 03 in that study. No California evaluation data is available for the new selections Muscat blanc 05 (Rauscedo VCR 3) and 06 (Milan, Italy). Muscat blanc ENTAV-INRA® 453 is available as a California registered selection.

Production

Production is usually 6 to 9 tons per acre. On good soils, yields of 10 to 12 tons per acre have been reported.

Harvest

Period: One of the earliest varieties, harvested in mid-August to mid-September.

Method: The short and occasionally woody peduncles make hand harvest difficult and require knives or shears for cluster removal. Canopy shaking results in medium-hard harvestability and heavy juicing. Trunk shaking results in medium harvestability and medium juicing. Fruit is mostly removed as single berries.

Training and Pruning

Muscat blanc is most commonly trained to bilateral cordons and pruned to 12 to 18 two-node spurs. Retaining low node numbers may limit cluster numbers unnecessarily and contribute to very compact clusters that are prone to bunch rot. Some very vigorous vineyards are head trained and cane pruned to assure adequate yield and to reduce bunch rot with more loose clusters. Mechanical hedge, non-selective pruning is an alternative to cane pruning.

leaves

Medium; mostly 3-lobed to almost entire with reduced lateral sinuses; closed U-shaped petiolar sinus; sharp, saw-like teeth; lower leaf surface glabrous.

shoot tips

Downy tips; young leaves bronze-red and shiny.

Trellising and Canopy Management

San Joaquin Valley vineyards can be trellised as a single-curtain with a single foliar support wire above the cordon wire. In coastal areas, vertical-shoot-positioned trellising would be appropriate for low to moderate vine vigor. The canopy tends to be moderately open because of its semi-erect shoot growth, with small to medium leaves and limited lateral shoot development. Most of the clusters will receive some direct sunlight in average canopies without leaf removal. Additional vertical foliar support and leaf removal may be warranted for bunch rot control. Excessive cluster exposure and sunburn can be a problem in weaker vines.

Insect and Disease Problems

Muscat blanc is very susceptible to powdery mildew, Botrytis bunch rot, and sour bunch rot. Bunch rot in new plantings may be lower with certain clones. On sandy soils, high nematode populations can further reduce vine vigor, especially on own-rooted plants.

Other Cultural Characteristics

Fully exposed berries will bronze or amber. Fruit will raisin if left on the vines past midseason. Fruit can be attractive to bees, wasps, and birds.

Winery Use

Muscat blanc is mostly used to produce quality, sweet, and light, muscat-type varietal wines, some of which are sparkling. These are commonly produced through cold fermentation and minimum skin contact. It is also used for dessert wines and to blend for added fruitiness.

—*L. Peter Christensen*

Muscat of Alexandria

Synonyms
Muscat d' Alexandrie and Muscat Romain are used in France, Moscatel Gordo and Moscatel Gordo blanco in Spain, Zibibbo and Moscatel romano in Italy, Muscat Gordo Blanco in Australia, and White Hanepoot in South Africa.

Source
Muscat of Alexandria is believed to have originated in North Africa and spread around the Mediterranean from the port of Alexandria, Egypt, possibly during the Roman Empire. It is widely known as a multipurpose variety; it is used as a table grape in Spain, Italy, Japan, and South America; a dessert wine and blending variety in southern Europe, Africa, South America, California, and Australia; a brandy (Pisco) variety in South America; as well as a raisin variety in the Old and New World.

The variety reached California in the mid 1800s, reportedly first brought in 1852 by Antoine Delmas, a member of the colony of French growers in Santa Clara County. It was also included in Agostin Harazthy's large variety importation from Europe in 1861. It became the dominant raisin variety in California until the early 1920s. These plantings added to the tonnage used by wineries for the production of muscat dessert wines when seedless varieties came to dominate the raisin market. Fresh shipment for home winemaking is another common outlet for California growers. Presently, acreage is fairly stable, after gradual losses from declining markets of muscat raisins and dessert wine.

Description
Clusters: large; long conical, can be winged; loose to straggly; relatively long peduncles.

Berries: large; oval, table grape-sized with pronounced muscat flavor; dull green to yellow when ripe, amber where exposed.

Leaves: medium to large; 3- to 5-lobed with narrow U-shaped petiolar sinus; large apical lobe; relatively sharp 2-ranked teeth; sparse tufted hair on lower surface; older leaves often have scattered patches of yellow tissue.

Shoot tips: downy white; young leaves yellow-green with bronze highlights.

Growth and Soil Adaptability
The vine is moderately vigorous to vigorous when grown on its own roots on medium- to fine-textured soils (sandy loam to clay loam); sandy soils cause very poor vigor. Shoot growth is semi-erect. Recommended in-row spacing is 6 or 7 feet and row middle spacing is 10 or 11 feet.

clusters
Large; long conical, can be winged; loose to straggly; relatively long peduncles.

berries
Large; oval, table grape-sized with pronounced muscat flavor; dull green to yellow when ripe, amber where exposed.

Rootstocks

Harmony and Freedom are successful in nematode-prone sites. They should be considered for use in sands to sandy loam soils due to Muscat of Alexandria's moderately high susceptibility to root knot nematode. These variety and rootstock combinations require close monitoring for potential zinc deficiency. A moderately vigorous to vigorous phylloxera rootstock should be used to assure adequate vine vigor and for leaf cover to avoid fruit sunburn.

Clones

FPS selections 02 and 03 (77- and 119-day heat treatments, respectively) are registered in the California certification program and have good fruiting characteristics. Selection 03 has been tested against two Australian selections, New South Wales J2 and G5. It demonstrated superior vine vigor, fruitfulness, and productivity with comparable fruit composition.

Production

Production is usually 7 to 10 tons per acre. Young, cordon-trained vines often produce more.

Harvest

Period: A late-season variety, harvested in early September to mid-October.

Method: By hand, harvest is easy to medium with medium to long, green cluster stems that are easily cut. Canopy shaking harvest is easy to medium; with light juicing. Trunk shaking is usually not the preferred method due to the low cordon heights of many vineyards.

Training and Pruning

The variety is most commonly trained to bilateral cordons and pruned to 12 to 18 spurs with one to two nodes each. The permanent vine framework should be fully formed before normal cropping to avoid overcropping and weak growth at the end of the cordon. Shoots and clusters should be thinned for crop adjustment through the training period. Cordon development over a 2- to 3-year training period may be required for vines of moderate vigor.

leaves

Medium to large; 3- to 5-lobed with narrow U-shaped petiolar sinus; large apical lobe; relatively sharp 2-ranked teeth; sparse tufted hair on lower leaf surface; older leaves often have scattered patches of yellow tissue.

shoot tips

Downy white; young leaves yellow-green with bronze highlights.

Head training is common in older vineyards; the vines are pruned to 8 to 20 one- and two-node spurs, depending on vine size. Vineyards trained to this system are usually preferred for fresh grape shipments. Many of the older, cordon-trained vineyards traditionally used cordon heights of 30 to 36 inches. New, cordon-trained vineyards should use cordon heights of 42 to 52 inches to facilitate machine harvest.

Trellising and Canopy Management

Exposed fruit is subject to sunburn, particularly during early summer hot spells. A foliage support trellis to shade fruit is recommended.

Insect and Disease Problems

Muscat of Alexandria is moderately susceptible to powdery mildew, black measles, and Pierce's disease and has low to moderate susceptibility for Phomopsis cane and leaf spot. Leaves may occasionally show blisters with densely hairy inner surfaces due to the feeding of the grape erineum mite. This tiny eriophyid mite is normally suppressed by sulfur application, but Muscat of Alexandria is more susceptible to the pest than most varieties. Leaves commonly show characteristic "muscat spot" symptoms in mid- to late summer. Symptoms appear as irregular, yellowish-chlorotic interveinal spots on the older leaves. Affected portions may also become necrotic or brown. This disorder is thought to be a genetic characteristic of some muscat varieties and is not related to any known disease, nutritional, or physiological problems. Certified, virus-free planting stock should always be used in new vineyard plantings due to leafroll-associated viruses in some existing vineyards.

Other Cultural Characteristics

The variety is susceptible to zinc deficiency, which results in poor fruit set and shot berries. This can be corrected by foliar spraying with neutral zinc or zinc oxide products before or during bloom. It is somewhat susceptible to overcropping. Exceptionally large yields—13 to 15 tons per acre—will shorten vineyard longevity, particularly in young, cordon-trained vineyards.

Winery Use

Muscat of Alexandria is used for muscat dessert, table, and sparkling wines, mostly in the warm districts of the San Joaquin Valley. It is also used in blending table and sparkling wines for the addition of fruity, muscat flavor and is a popular variety for fresh shipment for home winemaking.

—*L. Peter Christensen*

Nebbiolo

Synonyms
In Italy, the variety is known as Spanna, Picoultener, Picotendre, Prunent, and Chiavennasca.

Source
Nebbiolo is considered the premiere varietal of northwestern Italy. It is believed to have been cultivated in the Langhe district before the fourteenth century. The earliest documentation (1303) is provided in Pier de' Cresxenzi's *Ruralism Commodorum*. A very late ripener, Nebbiolo is named for the fog ("nebbia") that settles in the foothills during the late October harvest. In Italy, there are about 13,000 acres grown, with important plantings in the Piedmont region. There are also plantings of Nebbiolo in Australia, and in North and South America. In California, plantings are more prevalent in the Central Coast; they are scattered through the North Coast, Sierra foothills and the Central Valley.

Description
Clusters: medium to large; long cylindrical, compact; long peduncles.
Berries: medium round to short oval; purple but red-purple in sunlight.
Leaves: medium; deeply 5-lobed lobed with wide U-shaped petiolar sinus; lateral sinuses also deep; large apical lobe; sharp, multi-ranked teeth; moderately dense hair on lower leaf.

clusters
Medium to large; long cylindrical, compact; long peduncles.

berries
Medium round to short oval; purple but red-purple in sunlight.

Shoot tips: felty with pink edges; young leaves yellow-green with bronze-red patches; end of shoots with many small leaves and relatively long internodes.

Growth and Soil Adaptability
Nebbiolo grows in a wide range of soils in Italy. In California, it is a very vigorous cultivar, and it performs best on shallower, less fertile sites. In all cases, Nebbiolo is a fussy grape for climate requirements; the most suitable is a warm, continental climate, with low ripening month mean temperatures (approximately 63°F) and low temperature extremes. In California, frost-free sites are needed due to its early budbreak. The best sites should not be excessively hot, as Nebbiolo will shrivel in very hot temperatures. High-quality Nebbiolo vintages occur when the weather is dry during ripening.

Nebbiolo is a sprawling, vigorous, and open vine with low to medium basal bud fertility. Canes are slender and long, with strong tendrils that attach readily to wire, making shoot positioning and pruning difficult.

Rootstocks
In Italy, 420A and Kober 5BB are used. In California, Teleki 5C and SO4 are commonly used.

Clones
There are three principle groups of clones of Nebbiolo described in Italy: Lampia, Rosè, and Michet. The Michet clones are infected with grape fanleaf virus, and they are not registered or recommended for use. The

genetic profile of the Rosè group is different than Michet and Lampia. There is a great diversity of Nebbiolo clones available in Italy, and new clones are under development.

In California, certified Nebbiolo FPS selection 01 has been available for many years. Non-registered Nebbiolo FPS 02 (CVT 36), 03 (CVT 36), and 04 (CVT 230) are all from Torino, Italy. Nebbiolo Lampia FPS 01 from Torino is also registered. Nebbiolo FPS 07 (CN36) and FPS 08 (CVT 230), presently provisionally registered, should be available commercially in the near future.

Production

Vines are moderately productive, but the highest-quality wine occurs in a yield range of 3 to 5 tons per acre.

Harvest

Period: Nebbiolo is among the last varieties to ripen. In the North Coast, harvest occurs in mid- to late October.

Method: Harvest is primarily done by hand. Clusters are large and relatively easy to pick. With machine harvesting, a canopy shaker results in easy to medium harvest, with single berries and some cluster parts removed and medium juicing. Trunk shaking is easy to medium, with single berries, some cluster parts, and a few whole clusters removed. Juicing is light to medium.

Training and Pruning

Nebbiolo is vigorous, and its canes are long and trailing. It is often cane pruned due to low basal bud fertility. In warm districts spur pruning is effective. Canopy management consists of shoot positioning, top (and side) trimming, and sometimes leaf removal. Vines are usually cluster thinned.

leaves

Medium; deeply 5-lobed lobed with wide U-shaped petiolar sinus; lateral sinuses also deep; large apical lobe; sharp, multi-ranked teeth; moderately dense hair on lower leaf.

shoot tips

Felty with pink edges; young leaves yellow-green with bronze-red patches; end of shoots with many small leaves and relatively long internodes.

Trellising and Canopy Management

Vertical-shoot-positioned systems are widely used in both Italy and California. In Italy, vine density is 1,300 to 2,000 vines per acre, and the Guyot training system is used with one single cane of 10 to 12 buds per vine or 8 to 10 buds if virus presence reduces vigor. In California, vineyards are often planted 8 by 7 feet using VSP systems. A fruiting wire is set 30 to 40 inches high. Two moveable sets of foliage wires are used. Some growers are also experimenting with divided canopy systems as a way to control vigor and produce balanced vines.

Insect and Disease Problems

Nebbiolo is sensitive to powdery mildew. It has low susceptibility to bunch rot.

Other Cultural Characteristics

Nebbiolo is vigorous and sprawling, and it requires extra attention in shoot positioning.

Winery Use

Nebbiolo is notoriously difficult to grow and make into fine wine outside of its home region of Piedmont. Even so, Nebbiolo-based wines, acid and astringent in youth, evolve with maturation into some of the most well-structured, longest living, and richly scented wines. Historically, Nebbiolo's relatively high acids, pronounced tannins, and long cool fermentations (up to two months in wooden vats) meant that the wine would need cellaring for several years before being considered ready to drink. Today, shorter fermentations with less time on the skins plus the use of small oak barrels for aging have made Nebbiolo a more satisfying wine. The wines have large amounts of tannins, but color often is lacking. Controlling crop load and fermentation temperatures seems to help develop better wine color.

—*Glenn McGourty*

Pinot Family

Pinot noir appears to be genetically unstable and new clones, resulting from "point mutations" of this variety, have been selected by growers who were attracted to their unique fruit color or shoot growth. In Pinot noir vineyards, it is not uncommon to find one or more vines with a single shoot that has characteristics quite unlike the others on the same plant. Depending on the type of mutation that has occurred, these characteristics may or may not be maintained when buds from the shoot are used to propagate new vines. However, if all buds on the new vines display the same attributes that were present on the original shoot, then a new clone or variety is born.

Pinot blanc, Pinot gris, and Meunier are all descendants of Pinot noir. Each differs from its parent in various ways, most notably in fruit color, and in the case of Meunier, the copious amounts of white hairs on the shoot tips. These varieties differ in fruit flavor and wine aroma that sets them even further apart from Pinot noir.

Pinot noir

Synonyms

There are numerous synonyms for Pinot noir and each of its variants that have become known as varieties. This is to be expected given its seven centuries of regular cultivation in the Old World and subsequent movement to several other wine grape-growing countries. Many synonyms are out of use; however, it is not uncommon for Pinot noir to be known by several different names in various growing regions of France as well as in other countries. The list of synonyms given here is far from complete and there are variations of these names in use.

In France, Pinot noir is known as variations of the following: Pineau de Bourgoyne, Franc Pineau, Noirien, Franc Noirien, Salvagnin, Morillon, Auvernat, Auvernaut noir, Plant Doré, and Vert Doré. In Germany it is called Burgunder blauer, Blauer

Spätburgunder, Clävner, Blauer-Klävner, Schwarzer Riesling, Möhrchen, and Schwarzer Burgunder. In Italy, it is known as Pinot nera; in Austria, Blauer Nürnberger; and in Hungary, Nagyburgundi.

Source

Pinot noir is perhaps the oldest cultivated variety of the genus *Vitis*. It is thought to be the cultivated vine described by Roman authors in the first century. By the fourteenth century it was known by several names—including Pinot—in different growing regions in France.

Growth and Soil Adaptability

Pinot noir tends to be a moderate- to low-vigor variety when grafted onto rootstocks that do not have *vinifera* in their parentage. To meet fruit quality objectives, higher-vigor vines must be aggressively managed to control crop level. As a result, deep, fertile soils are usually not considered optimal for this variety. In California, Pinot noir is grown in a wide variety of soil types, from sandy loams to heavy clays.

Although soil type and depth will impact vine growth, climate will also affect growth and play a large role in determining site suitability for Pinot noir. To optimize fruit quality, cool areas are strongly preferred. Vines are spaced 4 to 6 feet apart in the vine row in most sites.

Pinot noir has among the earliest budbreak and harvest dates when compared to most varieties. Since it is a short-season variety, it is chosen for marginal sites where temperatures preclude other varieties from reaching full maturity. Early budbreak often puts it at risk for spring frost. Likewise, temperatures are more likely to be cool and damp during the bloom period, which can result in coulure or millerandage, thus reducing fruit set and yield.

Rootstocks

Several different rootstocks are used for Pinot noir, including those that tend to reduce vine growth as well as those that impart vigor. In extremely cool areas, rootstocks that may delay the development of ripe fruit characters, such as 110R, should not be used. Occasionally, virus-infected budwood will result in severely diseased vines when grafted onto specific rootstocks. Rootstock selection is a function of the site's soil and climate as well as production goals.

Clones

There are more clones of Pinot noir than of any other wine grape variety, and, not surprisingly, most are from France. California nurseries that have obtained registered budwood from FPS offer a tremendous diversity of Pinot noir selections. Even more selections are being registered for the certification program pending disease therapy and testing. At present nearly 100 Pinot noir selections have been submitted to UC Davis for inclusion in the registration program, including French clones and heritage California selections. More than half of these are commercially available as FPS-registered selections. Many of the most respected Pinot noir clones were developed by programs in Dijon, France, and are commonly

known as "Dijon clones" in California. The most authentic Dijon clones are available through the ENTAV-INRA® trademark program. These include Pinot noir ENTAV-INRA® 115, 165, 236, 375, 459, 667, 743, 777, and 943. There are also Dijon clones of varied source and disease status in the trade.

The viticultural performance of Pinot noir clones as well as wine quality evaluations often produce inconsistent results for the same clones due to differences in site, climate, and management practices. Also, growers have access to clones that have not been planted in California previously and for which there is little, if any, performance information. This makes it extremely difficult to do more than generalize about the behavior of specific clones. There are four main groups of Pinot noir clones: standard quality (Pinot fin), highly fruitful (Pinot fructifier), upright shoots (Pinot droit), and loose-clustered (Mariafeld). Past experience has shown that excellent quality wines have been produced from clones within each group, although they often do not achieve this reputation in France.

Wine-making goals are the driving force behind clonal selection in Pinot noir. In extremely cool areas, early ripening is a necessity, hence lower-yielding clones would be chosen over high-yielding, later-maturing clones. In warmer areas, moderate-yielding clones can be successful. Practical considerations such as managing Botrytis bunch rot may cause growers to consider clones that reportedly tend to have looser clusters, such as FPS selections 17 and 23. However, experience has shown that in areas with high rot potential, clonal differences in cluster architecture seldom significantly reduce disease severity.

Pinot noir clones affect vine fruitfulness primarily by impacting the number of berries per cluster, although they also affect berry weight and cluster number. Due to the different aptitudes of the Pinot noir clones, many growers prefer to plant more than one clone of this variety in a vineyard. Higher-producing clones such as FPS selections 31 (French 236), 32 (French 386), and 33 (French 388) are commonly used for sparkling wine production since they are harvested at lower sugar maturity. There are several more clones from France's Champagne region that have been introduced to California recently.

Several clones are used in coastal vineyards for table wine production. These include (but are not limited to) ENTAV 114, 115, 667, and 777 as well as FPS selections 04 and 05, both known as Pommard. Field selections have produced high-quality fruit and wines when located in optimum sites. These are commonly known by grower names such as Swan and Martini. Single-vine accessions of these field selections and others from selected California vineyards are in the registration process at FPS.

Production

Pinot noir tends to be fruitful and can overcrop itself for a specific wine quality target. It is thought that large crops preclude ultra-premium wine production. Depending on vine density, 3 to 5 tons per acre is acceptable for table wine production. Higher yields usually result in quite acceptable, yet unremarkable red wines. Tonnage for sparkling wine production may be the same or it may be slightly higher depending on the winemaker's preference.

Weather conditions at budbreak and bloom will affect set. In warm years, most clones may set too much fruit, and cluster thinning would be required to reach desired yields. Because Pinot noir vines produce shoots from latent buds, early season shoot thinning is required to achieve one shoot per count bud. Depending on the vines and management style, shoots may be thinned a second time prior to bloom to re-establish optimum shoot number. Clusters may be thinned to a prescribed method during the period of time that they are acquiring color. Cluster thinning may occur again near the end of veraison. For sparkling wine production, shoot thinning will occur only once, and vines may not be cluster thinned.

Cool and rainy weather during bloom can reduce fruit set dramatically. It is not uncommon for yields to be half the average under these conditions. Low-yielding clones exacerbate this effect.

Harvest

Period: An early season variety, harvested in August to early September for sparkling wine and throughout September for table wine.

Method: Pinot noir is often harvested by machine for sparkling wine production; however, it is also hand harvested and transported to the winery in small tubs or half-ton bins. When machine picked for sparkling wine, it is essential to press the grapes soon after harvest to reduce color in the must. Harvesting Pinot noir grapes for table wine is done by hand for ultra-premium production. It is also machine harvested. Harvestability is easy to medium, with single berries and some cluster parts removed. Juicing is light to medium, with less juicing with bow rods than straight rods. Trunk shaking harvestability is medium, with light juicing.

Training and Pruning

Pinot noir vines are spur pruned and cordon trained, or they may be head trained with fruiting canes and renewal spurs. In low-vigor sites, the Guyot training system of one cane and one spur is sometimes preferred to maximize uniform maturity by minimizing the amount of fruit arising from a large number of renewal spurs. In small-cluster clones or in very cool production areas, cane pruning is often practiced. In more vigorous sites, shoots will be hedged as needed.

Trellising and Canopy Management

The height of the fruiting wire for a vertical-shoot-positioned system is commonly 30 to 36 inches. Two pairs of moveable wires are used to position shoots vertically. In vigorous sites, lyre or U-shaped horizontally divided canopies are used. GDC should be avoided since fruit tends to sunburn. Depending on row orientation, vertically separated or divided canopies such as Scott Henry or Smart-Dyson may result in over-exposed fruit without leaf removal. Leaf removal is practiced to enhance air movement around the fruit and reduce the severity of Botrytis bunch rot. Often leaf removal is minimal to avoid sunburn.

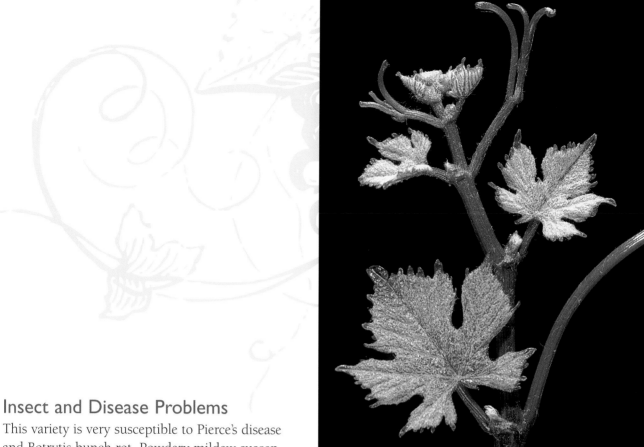

Insect and Disease Problems

This variety is very susceptible to Pierce's disease and Botrytis bunch rot. Powdery mildew susceptibility is a result of the mild temperatures common in the regions where the variety is grown. Early budbreak may make it more susceptible to thrips.

Other Cultural Characteristics

Cool conditions just prior to bloom may cause small necrotic areas to form at the attachment point of the petiole to the blade in the distal leaves of some shoots. The blades will abscise and lateral shoots will push at those nodes, which gives the affected shoot a bushy appearance. Occasionally, the cluster stem will also become necrotic in one spot, and often this will cause the entire cluster to desiccate. This variety is extremely susceptible to coulure when temperatures are low just prior to bloom.

A large number of lateral shoots is common. A significant second crop (1 to 1.5 tons per acre) may develop. Depending on the season, vines harvested for sparkling wine may mature second crop clusters by early fall, allowing growers the option to harvest them for table wine if the price warrants.

Winery Use

Pinot noir may be harvested at 18 to 20° Brix to produce a sparkling wine that is usually white. For red table wine, grapes are harvested beginning at 23.5° Brix. The wines usually do not have an intense color even in cool areas; however, they are known for their aroma and flavor under these conditions. When grown in hot areas, both color and flavor are reduced.

—Rhonda J. Smith

Pinot Blanc

Like the other pinots, there are several synonyms for Pinot blanc. In Germany, it is known as Weisser Burgunder, Clevner, Clävner, Weisser Ruländer, and Weisser Arbst. The most commonly used name in Italy is Pinot Bianco, and in Austria it is Weissburgunder. Beli Pinot, Feherburgundi, and Rouci Bile are synonyms used in Eastern Europe.

Pinot blanc is a mutation of Pinot noir that resulted in white grapes. It made its first appearance in Europe later than Pinot noir and Pinot gris, and it is speculated that Pinot blanc was in the Alsace in the sixteenth century. In the 1980s, most Pinot blanc vineyards in California were discovered to actually be the variety Melon. In other countries, Chardonnay and Chenin blanc were misidentified as Pinot blanc.

Pinot blanc is much easier to grow than Pinot noir. It has a small canopy yet is relatively fruitful, thus cane pruning may not be necessary. On a six-foot vine spacing, it will produce no more than about 12 pounds of fruit per vine. A wider vine spacing would not be used. The clusters are extremely tight—as much if not more so than all of the other members of the Pinot family. In coastal areas, Pinot blanc for still wine will ripen during the first three weeks of September.

Oregon growers have evaluated two clones from Colmar that they refer to as ENTAV 55 (formerly labeled INRA 159) and ENTAV 54 (formerly labeled INRA 161), which make up the plantings there; these are "generic clones" rather than trademarked clones. FPS Pinot blanc 07 is a generic clone, also derived from French 55. Also registered are Pinot blanc FPS 05 and 06, which originated from Rauscedo, Italy. ENTAV-INRA® 54 is available commercially and registered at FPS. Unlike Pinot gris, Pinot blanc is planted almost exclusively in cool growing regions.

A white varietal wine is made from Pinot blanc grapes. Pinot blanc has been blended with other varieties to produce sparkling wines.

—Rhonda J. Smith

Pinot Gris

Pinot gris is known by several different names throughout its growing regions in Europe. It is called Pinot burot, Gris Cordelier, and Malvoisie in France. In Germany it is known as Ruländer and Grauer Burgunder, and in Switzerland as Malvoisie. The Italians call it Pinot grigio, and the Hungarians Szürkebarat.

Pinot gris, like Pinot noir, was first described in the fourteenth century. It is widely planted throughout Europe; in the United States, Oregon growers were early pioneers in planting and promoting the variety.

Pinot gris is a variant of Pinot noir that produces a grape with variable color described as pinkish, coppery gray, and brownish pink. In the vineyard, it is not uncommon to find clusters that contain one to several berries that are darkly colored and others that are white. Also, individual berries may have skins that display the entire range of colors. A white wine is made from this variety.

Even more so than Pinot noir, Pinot gris requires a cool climate and a long growing season in order to maintain its slightly low acidity. To make a wine style that is typical of cool growing regions, it is harvested at no more than 23.5° Brix. In the coolest growing regions, harvest sugar level is often determined by the year since extremely cool vintages may result in fruit that is less ripe. This variety is also grown in warmer areas such as the Sierra foothills and the northern San Joaquin Valley, where it ripens easily.

Vineyard design, including rootstock, trellis, and planting density is similar to that of Pinot noir. Vigor is considered moderate to moderately low, although this is affected by site characteristics such as soil depth and temperatures. It may be slightly less vigorous than Pinot noir. For premium wine quality, yields are kept under 5 tons per acre. In cool areas, low tonnage often occurs by default, since cool, damp conditions at bloom will result in poor set.

Pinot gris is more versatile and easier to grow than Pinot noir. In warmer areas, it is acceptable to plant it on deeper soils that are capable of producing up to 6 tons per acre. In these areas, cluster thinning is not as essential. Fruit set is variable depending on the year, yet tight clusters are the norm.

There are three registered selections of Pinot gris. Registered FPS 01 is from the University of California's Jackson Vineyard. Registered FPS 04 and 05 are both from generic French clone 53 that came to FPS from Oregon. Several additional Pinot gris selections are being treated, tested, and identified to qualify them for registered status, including FPS 11 (old selection from Alsace, France); 06 (from Germany); 09 (French 52); 08 (VCR 5 from Italy); and 10 (SMA 505 from Italy). No information is available on the relative performance of these selections.

—*Rhonda J. Smith*

Pinot Meunier

The variety is known by different names in France, including simply Meunier. In Germany, it is called Schwarzriesling, and in Australia it is known as Miller's Burgundy for the floury-white appearance of the shoots.

Pinot Meunier is the most distinct variant of Pinot noir in terms of physical appearance and growth habits. It is a red grape variety with pendulant shoots that are covered with copious amounts of white hairs that increase in density toward the terminal ends so that shoot tips appear nearly white.

Pinot Meunier is a moderately high-vigor variety that can be planted in unrestricted soil. Its pendulant canes preclude tight vine spacing and require multiple passes to position shoots. In vertical canopies, shoots are stuffed between three pairs of stationary wires. Several different rootstocks may be used; however, those that impart vigor may require training to a divided canopy. For either cordon- or head-trained, cane-pruned canopies, cluster exposure should be limited.

Clusters are small and extremely tight. In high-yielding years, there is an increase in the number of winged clusters. There are very few clones of Pinot Meunier, and there is little difference in fruitfulness among planted areas. Budbreak is about the same time or slightly later than Pinot noir, yet it is harvested at the same time or slightly earlier than Pinot noir for sparkling wine. The variety can be easily harvested by machine; however, depending on wine style, it may be hand harvested in small tubs since the juice oxidizes quickly. A second crop is consistently produced and occasionally harvested for still wine two to three weeks after the primary clusters are picked for sparkling wine.

Production goals are usually higher for this variety than for Pinot noir, and acceptable yields can be as high as 8 tons per acre. In California it is used as a component of sparkling wine and, to a much lesser degree, it is made into a varietal red table wine.

In cool springs, portions of the cluster will become necrotic and desiccate, just as in Pinot noir. Pinot Meunier is not as susceptible to powdery mildew as Pinot noir, and it is far less susceptible to Eutypa dieback than Chardonnay. Although it has tight clusters, Botrytis bunch rot is not often an issue since Pinot Meunier is harvested before fall rains.

There are now six selections of Pinot Meunier in the collection at UC Davis: FPS 01 (French 817), 02 (French 864), 03 (French 791), 04 (French 819), 05 (French 864), and 06 (French 819). Of these, FPS 01, 05, and 06 are registered. There is no information available on the relative performance of these selections.

—*Rhonda J. Smith*

Riesling

Synonyms

Riesling or White Riesling are the approved synonyms. In the United States the name Johannesburg Riesling has been used to distinguish the variety from other non-Riesling varieties such as Grey Riesling (whose correct name is Trousseau gris) and Emerald Riesling (a Riesling × Muscadelle hybrid).

Source

Riesling is the noble wine grape variety of Germany, where, in the Rhine and Moselle regions, the grapes have produced distinctive, quality wines for centuries. The cultivation of Riesling in Germany is believed to date back to the time of Roman occupation. It has been grown in California since the late 1800s. It is most commonly grown in the cooler production regions, with the majority of the acreage found in the Central Coast.

Description

Clusters: small; cylindrical to globular, can be winged, compact; short peduncles.

Berries: small; round; white-green with prominent lenticels.

Leaves: small; 3- to 5-lobed and often overlapping to appear entire; closed U-shaped petiolar sinus, purple petioles and nodes on growing shoots; leaf surface bullate; short, rounded teeth; light scattered hair on lower leaf surface.

Shoot tips: downy, green-white tips; young leaves yellowish with bronze-red patches.

Growth and Soil Adaptability

Vine vegetative growth can vary significantly from weak to moderately vigorous depending on the climatic region, soil characteristics, moisture availability, and rootstock selection. Adaptable to a wide range of soil types, the vine's highest vigor will be on fertile soils with high-moisture availability. Shoot growth on non-positioned canopies is fairly upright, but vines develop long, trailing shoots when growth is vigorous. Vine in-row spacing can vary from 4 to 6 feet.

Rootstocks

Riesling has no known incompatibilities when certified budwood is used to propagate the planting stock. Rootstock selection should be based on the pest situation, soil characteristics, and potential vine vigor of the site. In the coastal areas Riesling has been successfully grown with the rootstocks Teleki 5C, Kober 5BB, 3309C, 110R, and Freedom. Rootstock experience is limited due to the low acreage replanted to the variety in the late 1980s and 1990s. Rootstocks may have more influence on sites where anticipated vigor is low, and the choice may have a greater effect on vine growth and development.

Clones

The clonal selection of Riesling began in 1921 at the Geisenheim Research Center in Germany. The goal was to preserve the genetic diversity and to select for consistent

clusters

Small; cylindrical to globular, can be winged, compact; short peduncles.

berries

Small; round; white-green with prominent lenticels.

yield and high wine quality. In Germany, clonal selection has been used to promote higher production and evenness in the vineyard. Many there believe that to produce quality wines there should be a blend of clones to increase complexity of flavors. In Germany, the Geisenheim selections 24Gm, 64Gm, and 94Gm are noted for light fruitfulness and good balance between all flavor components. Geisenheim selection 110Gm has an extremely fruity, slightly muscat flavor, and in warmer sites it is regarded as not typical of German Riesling wines. Geisenheim selection 198Gm has lower crop yields with wines of elegant fruitfulness and pronounced flavor, but with all components in good balance. Geisenheim selection 239Gm is the most widely distributed selection in Germany and produces wines with a range of turpenes, resulting in a spectrum of fruitfulness.

In California, White Riesling FPS 02 is sourced from the Geisenheim selection 198Gm. White Riesling FPS 03 and 09 are from Geisenheim selection 110Gm. White Riesling FPS 12 originates from clone 90 from Neustadt, Germany. Other registered White Riesling selections include FPS 04 from an unknown source, and FPS 10 from the Martini Vineyard.

Production

Vine yield can vary considerably by climatic region, site influences, and cultural practices. Crop size can range from 4 to 8 tons per acre. Riesling tends to overcrop when it is grown on deep, fertile sites.

Harvest

Period: A midseason variety, harvested in mid-September in the warmer areas and mid- to late October in the cooler production areas.

Method: Hand harvest is easy using knives or shears. Horizontal rod or bow machine harvest is easy to moderate with fruit coming off as single berries with moderate juicing. Bow-rod picking heads used on well-trained vines on vertical-shoot-positioned trellises have lower shoot and spur breakage than straight rods. Trunk shaker machine harvest is easy with fruit coming off as single berries and light to medium juicing.

leaves

Small; 3- to 5-lobed and often overlapping to appear entire; closed U-shaped petiolar sinus, purple petioles and nodes on growing shoots; leaf surface bullate; short, rounded teeth; light scattered hair on lower leaf surface.

shoot tips

Downy, green-white tips; young leaves yellowish with bronze-red patches.

Training and Pruning

Due to the small cluster size the variety was traditionally head trained and cane pruned. Cordon training with spur pruning has been used successfully in many areas. In vineyards where bud fruitfulness is low, cane pruning may result in higher production.

Trellising and Canopy Management

For low- to moderate-vigor sites the use of vertical-shoot-positioned systems is most appropriate. For high-vigor sites the use of Smart-Dyson or Scott Henry for vertical systems or the use of a horizontal (GDC or lyre) system may reduce canopy shade.

Insect and Disease Problems

The fruit is highly susceptible to infection by *Botrytis cinerea.* Pre-harvest rain can cause high levels of bunch rot at harvest.

Other Cultural Characteristics

In some seasons blind buds may occur in the mid-cane area. Riesling leafs out approximately one week after Pinot noir.

Winery Use

Riesling can produce table wines that are distinctive in aroma and flavor. The wines can have intense fruit aromas of apricot or peach. Wine styles range from dry to very sweet dessert wines. The highest quality is achieved when the grapes are grown in the cooler production areas. It is well suited for the production of late-harvest dessert wines. The susceptibility to Botrytis infection and the retention of acidity through the very late stages of ripening allow the grapes to become concentrated by dehydration and still retain sufficient acidity to balance the high residual sugar.

—*Larry J. Bettiga*

Roussanne

Synonyms

The variety is known in France as Barbin, Bergeron, Martin Cot, Fromental, Fromental jaune, Fromenteua, Rebolot, Babellot, Ramoulette, Remoulette, Greffou, Picotin blanc, and Courtoisie.

Source

Roussanne is from the Rhône Valley region of France. Roussanne is thought to have originated in the middle Rhône and Isere valleys. In the mid-twentieth century, as Roussanne vineyards became infected with fanleaf virus, the higher-yielding Marsanne was frequently replanted. Cleaner budwood sources resulted in the planting of Roussanne to newer areas in the Cotes du Rhône (south), where it is frequently blended with other varieties for white wine. It is also planted in Lucca, Italy, and Australia. In California, there were 194 acres in 2000, mostly planted in the Central and North Coast districts.

Description

Clusters: medium; long cylindrical with moderate shoulders, well-filled to compact; medium-length peduncles

Berries: small; round; yellow-amber.

Leaves: medium; deeply 5-lobed with reduced inferior lobes that appear like shoulders; closed, narrow U-shaped petiolar sinus; distinct bullate surface; broad rounded teeth; glabrous to sparse tufted hair.

Shoot tips: felty tips; young leaves yellow and downy.

Growth and Soil Adaptability

Roussanne is moderate in growth. Canes are slender and somewhat pendant. Clusters are small. In California, Roussanne is considered an early variety in budbreak and ripening. Roussanne breaks bud and ripens about a week after Chardonnay and Viognier, making it an early to midseason variety. Roussanne needs a sunny, dry climate to ripen fruit. It is grown in the northern Rhône Valley, with a climate similar to low Winkler Region III. In France, Roussanne does best on well-drained, low fertility soils of calcareous parent material. In California, most vineyards are planted on fertile alluvial soils.

Rootstocks

In France, 110R and 41B are used. In the United States, rootstock experience is limited, but Teleki 5C, SO4, and 3309C rootstocks are used.

Clones

There are two clones of Roussanne in France, ENTAV 468 and ENTAV 522. Both were developed primarily to ensure that there was a clean source of budwood not infected with fanleaf. Both are reported to produce well with good quality fruit. Roussanne ENTAV-INRA® 468 is available as California certified stock. In addition, a California vineyard selection Rousanne FPS 02, now available to nurseries, should be available to growers in the near future.

clusters

Medium; long cylindrical with moderate shoulders, well-filled to compact; medium-length peduncles.

berries

Small; round; yellow-amber.

Production

In the Rhône region, it is a moderate bearer, yielding less than Marsanne but more than Viognier. Maximum authorized yield is about 3 tons per acre. In the United States, Roussanne has been misidentified and may be Viognier in some of the early plantings. Yields are low to moderate, in the 3 to 5 ton per acre range.

Harvest

Period: An early to midseason variety.
Method: Most California vineyards are hand harvested.

Training and Pruning

Roussanne is pruned goblet style in many Rhône vineyards or on simple vertical-shoot-positioned trellis systems. In most appellations, vines are planted in very high density (varying from 907 to 1,361 vines per acre). Pruning ranges from a low of three buds per vine to a maximum of six spurs of two buds. In California, most vineyards are cordon trained on vertical-shoot-position trellises and spur pruned. Spurs are closely spaced (5 to 7 inches apart) using two-node spurs ranging from 36 to 56 buds depending on vigor of the vine and spacing. Newer vineyards are being planted 4 to 6 feet apart in rows 8 feet apart.

Trellising and Canopy Management

For moderate vigor, most plantings are trained on vertical-shoot-positioned systems. Fruiting wires are set between 30 to 36 inches in height. Shoots are positioned upright between two sets of movable foliage wires. Leaves are pulled lightly around the clusters for good air circulation.

leaves

Medium; deeply 5-lobed with reduced inferior lobes that appear like shoulders; closed, narrow U-shaped petiolar sinus; distinct bullate surface; broad rounded teeth; glabrous to sparse tufted hair.

Felty tips; young leaves yellow and downy.

Insect and Disease Problems

In France, many of the northern Rhône Roussanne vineyards suffered from fanleaf virus and were subsequently replanted to the higher-producing Marsanne. Clean nursery stock programs have solved that problem. Roussanne is also quite susceptible to powdery mildew. Botrytis bunch rot can also be a problem, especially in years when pre-harvest rains occur.

Winery Use

In France, Roussanne is usually blended with Marsanne and vinified as either a still or sparkling wine. The flavor descriptors include floral, honey, apricot, and sometimes mineral, steel, or lean, which are devoid of fruit or yeast flavors. Roussanne has good acidity, good tannic structure, and a strong floral component when carefully vinified. Wines are generally fermented in stainless steel tanks, and are often bottled directly without barrel contact. Other producers age the wine for limited time in new oak barrels. Wines are usually consumed while young (less than 4 years old). Roussanne is fresh and lively compared to Marsanne, which tends to oxidize more easily. Carefully made Roussanne with good acidity can also age well. Roussanne is also blended sometimes with Syrah in the northern Rhône Valley to soften the tannins and intensity of those frequently powerful red wines.

—*Glenn McGourty*

Rubired

Synonyms
None

Source
Rubired is a hybrid released by H.P. Olmo of the University of California, Davis, in 1958. It was produced by crossing Alicante Ganzin and Tinta Cao. This is a teinturier variety; its berries have both red flesh and juice. Alicante Ganzin is a French hybrid created by crossing Aramon Rupestris Ganzin #4 and Alicante Bouschet, and is used primarily for breeding teinturier cultivars. Tinta Cao is a distinguished variety from northern Portugal used to make premium port wines.

Description
Clusters: medium to large; well-filled to compact; medium peduncles.
Berries: small; round; purple with deep purple-red juice.
Leaves: medium; mostly entire; narrow U-shaped petiolar sinus; sharp, short teeth; upper surface glossy and waxy; leaf underside glabrous; leaf margin rolled under.
Shoot tips: cobwebby tip; young leaves glabrous with strong red pigmentation in shoots as well.

Growth and Soil Adaptability
Rubired is adapted to a wide variety of soils in the San Joaquin Valley. Own-rooted vines grow vigorously in loams and loamy sands. Lower vigor may be expected on coarse, sandy soils. Rubired is more vigorous than Barbera and less vigorous than Colombard. Its growth habit is semi-upright and open. In-row spacing generally ranges from 7 to 8 feet for own-rooted vines.

Rootstocks
Most vines in the San Joaquin Valley are planted on their own roots. Own-rooted vines exhibit some tolerance to root knot nematodes once roots are 5 months or older. Own-rooted vines are highly susceptible to the fanleaf virus host *Xiphinema index* nematode in coarse-textured, sandy soils. Freedom, Ramsey, or Harmony may be used where nematodes pose a problem.

Clones
Since this variety is relatively new, field selections differing in viticultural performance or growth characteristics have not yet been identified. Growers should plant only virus-tested certified stock; many of the original plantings were infected with virus diseases, particularly leafroll. Two registered selections, FPS 02 and 03, are available.

Production
Production in the San Joaquin Valley generally ranges between 8 and 12 tons per acre.

clusters
Medium to large; well-filled to compact; medium peduncles.

berries
Small; round; purple with deep purple-red juice.

Harvest

Period: A late-season ripening variety, typically harvested from early to late-September in the San Joaquin Valley.

Method: While it is easy to remove from the vine, Rubired is not popular with harvest crews due to its numerous and relatively light-weight clusters. It is easy to machine harvest with a canopy shaker, with most fruit removed as single berries and moderate juicing. Trunk shaking is also easy, with medium juicing and most fruit removed as single berries. Young vines are generally easier to harvest than mature vines.

Training and Pruning

Rubired is trained to bilateral cordons and pruned to 14 to 16 two- to three-node spurs per vine. Basal buds are usually quite fruitful, making young vines susceptible to overcropping. Shoot thinning of 3- to 4-year-old vines prior to bloom is recommended to reduce excessive crop, decrease the number of straggly clusters per vine, and reduce canopy density. Mature vines with large crop loads tend to set straggly clusters. This facilitates the use of machine-hedge pruning and the retention of high node numbers on large, vigorous vines.

Trellising and Canopy Management

Rubired is commonly trellised to the traditional California two-wire vertical system.

Insect and Disease Problems

Fruit and foliage are moderately resistant to infection by powdery mildew. Fruit susceptibility to bunch rot is very low, which allows late harvest. Rubired is very susceptible to Eutypa dieback. Young vines are very susceptible to collar rot, thus the area near the trunk should be kept free from standing water or water-saturated soil. It is highly susceptible to spider mite infestations.

leaves

Medium; mostly entire; narrow U-shaped petiolar sinus; sharp, short teeth; upper surface glossy and waxy; leaf underside glabrous; leaf margin rolled under.

shoot tips

Cobwebby tip; young leaves glabrous with strong red pigmentation in shoots as well.

Other Cultural Characteristics

Vine stress due to overcropping, insect damage, or other factors may result in delayed, erratic budbreak in the spring. This may cause the vines to enter an alternate bearing pattern, in which yields fluctuate drastically from year to year. Vine stress from overcropping also leads to poor wood maturity, increasing the likelihood of injury or vine damage due to exposure to cold temperatures during the winter. Berry shrivel may be a problem in over-ripe fruit, and yield can drop 30 percent or more in 10 to 14 days. The variety suckers readily at base of trunk.

Winery Use

Rubired is used primarily for red juice concentrate production, commonly utilized for blending purposes in the winery, as well as for food products including fruit juices. It produces a dark red blending wine, with little character or body, and is used to increase the color of generic or varietal table and dessert wines.

—Nick K. Dokoozlian

Ruby Cabernet

Synonyms
None

Source
Ruby Cabernet is the resulting cross of Carignane and Cabernet Sauvignon made by H.P. Olmo of the University of California, Davis. It is the oldest crossing of wine varieties made by Olmo, released for distribution in 1948.

Description
Clusters: medium to large; long conical, winged, loose to well-filled clusters; medium to long peduncles.

Berries: medium; round to short oval; dark purple with distinct Cabernet Sauvignon flavor.

Leaves: large; floppy, deeply 5-lobed leaves; closed U-shaped petiolar sinus; often with teeth in superior lateral sinuses; long, sharp teeth; lower leaf surface glabrous; pink petioles and veins; slight to moderate puckering on leaf surface reminiscent of Carignane.

Shoot tips: downy; young leaves downy and bronze-red.

Growth and Soil Adaptability
Vines have high vigor on deep, fine, sandy loam to clay loam soils. Vigor is low to moderate on very coarse sands. The canopy tends to be open and shoot growth semi-erect. Vine spacing should be about 7 to 8 feet or more down the vine row for vertical-shoot positioning or standard bilateral cordons. For horizontally divided quadrilateral vines, spacing should be 6 to 7 feet.

Rootstocks
Moderate vigor rootstocks should be used, such as 101-14 Mgt or Kober 5BB, although very sandy soils or shallow soils may require more vigorous rootstocks such as Freedom, 1103P, or possibly 110R. Machine hedging for pruning or possibly minimal pruning may produce acceptable quality for less expense. Own-rooted vines or vines grafted on Freedom have done well, although both choices entail long-term risk, especially own-rooted vines, due to potential soil pest susceptibilities.

Clones
This variety, introduced in 1948, was too recent for true clonal diversity to have arisen, so there are no true Ruby Cabernet clones. There are two registered selections of Ruby Cabernet: FPS 02 and 03. These selections differ only from the original mother vine by the number of heat-treatment days. They can be expected to perform identically.

clusters

Medium to large; long conical, winged, loose to well-filled clusters; medium to long peduncles.

berries

Medium; round to short oval; dark purple with distinct Cabernet Sauvignon flavor.

Production

Ruby Cabernet vines are very productive, capable of bearing large crops—8 to 12 tons per acre. Higher yields may sacrifice color intensity and delay harvest.

Harvest

Period: A mid- to late-season variety, harvested in mid-September to mid-October.

Method: Hand harvest is very difficult due to the thick, woody cluster stem. Vines are relatively well suited to trunk shaker harvesters, with medium difficulty of fruit removal. Berry juicing potential at harvest is light to medium. Machine harvest by pivotal striker works fairly well as fruit detaches mostly as berries, but newer canopy shaker heads may reduce juice loss and vine damage.

Training and Pruning

The variety tends to overproduce during vine training due to its short to medium node spacing and its high fruitfulness. Shoot and cluster thinning are essential to adjust crop load and develop strong spur positions during the early years of cordon development. The variety is well suited to spur pruning and bilateral cordon or quadrilateral training. A spur count of 16 to 22 two-node spurs is acceptable, depending on rootstock, soil depth, and soil texture. It is also adapted to mechanical hedge pruning ("box pruning") on bilateral cordons at a traditional vine spacing of 7 to 8 feet.

Trellising and Canopy Management

One or two foliage wires or a two-wire "T" trellis would be acceptable for most conditions. A cordon system with no foliage wires is beneficial for vines intended for machine pruning and harvesting. Vertical-shoot-positioned systems or head-trained vines are possible, but not economically advantageous. There is little need for shoot thinning or especially leaf removal (in most situations) with regard to quality and disease concerns.

leaves

Large; floppy, deeply 5-lobed leaves; closed U-shaped petiolar sinus; often with teeth in superior lateral sinuses; long, sharp teeth; lower leaf surface glabrous; pink petioles and veins; slight to moderate puckering on leaf surface reminiscent of Carignane.

Insect and Disease Problems

Ruby Cabernet is moderately sensitive to powdery mildew and very resistant to Botrytis and sour rot. It shows some tolerance to Pierce's disease.

Other Cultural Characteristics

Budbreak ranks among the later group of commercial varieties. Berries can sometimes set irregularly, especially on sandy, zinc-deficient soils. Fruit holds well on the vine during the late stages of maturity.

Winery Use

Ruby Cabernet produces quality wines with good color under optimum conditions, especially at lower crop levels. The wine can have distinctive varietal character. Potential wines can be used for less expensive varietals or as a blend to extend Cabernet or claret-type wines. Some interest in the variety as a premium wine for warmer regions may increase.

—*Paul S. Verdegaal*

Sangiovese

Synonyms

Clonal diversity and geographical designations have resulted in a profusion of synonyms in Italy. The two main sub-types are Sangiovese Grosso and Sangiovese Piccolo. Sangiovese Grosso, also known as dolce and gentile, is synonymous with Sangiovese di Lamole of the Greve-Firenze region of Chianti, the Prugnolo gentile of Montepulciano, and the Brunello of Montalcino. Sangiovese Piccolo, also known as forte and montanino, is synonymous with Sangiovese di Romagna, Sangiovese del Verrucchio, and Sangioveto. More recently, Italians are questioning the entire Sangiovese Grosso/Piccolo designation and believe that the diversity of the cultivar makes a separation of the two groups questionable.

Source

Sangiovese is the most planted variety in Italy (about 10 percent of total acreage), especially in Tuscany, where it is thought to have originated, with documentation as far back as the sixteenth century. It is also grown extensively in the Mendoza province in Argentina, increasingly in California, and to a limited extent in Australia. Plantings in California date back to about 1880.

Description

Clusters: medium; wide and long conical, well-filled; long peduncles.

Berries: medium; oval; blue-black. Many clonal types including "piccolo" types with very small berries; some berries are seedless due to fanleaf virus infection.

Leaves: large; 3-lobed with large, triangular apical lobe; open U-shaped petiolar sinus; narrow lateral sinuses, inferior very shallow; short, sharp teeth; lower leaf surface mostly glabrous with scattered tufted hair.

Shoot tips: felty; young leaves green with yellow-bronze highlights.

Growth and Soil Adaptability

The vine is of medium-high vigor, and growth is semi-upright to trailing, with long, strong shoots. Recommended in-row spacing is 6 or 7 feet on high-vigor sites and 4 or 5 feet on low-vigor sites. Wineries will more often use Sangiovese on shallow, more limited soils to avoid high vigor that can be detrimental to wine quality. It is adapted to cool to warm climate regions.

Rootstocks

Various rootstocks have been successfully used in California, including Teleki 5C, SO4, Kober 5BB, 420A, 110R, 3309C, 101-14 Mgt, and St. George. High-vigor stocks such as 140Ru or Freedom should be used with caution on fertile soils and when needed for lime or nematode tolerance.

Clones

Clonal diversity in Italy is great, with at least 35 clones from different regions registered in the National Catalogue. Sangiovese FPS 02, 03, and 04 have shown significant

clusters

Medium; wide and long conical, well-filled; long peduncles.

berries

Medium; oval; blue-black. Many clonal types including "piccolo" types with very small berries; some berries are seedless due to fanleaf virus infection.

differences in fruit characteristics and composition in vineyard trials replicated in California. Sangiovese FPS 02 is the most fruitful, sometimes with smaller berries. However, it may require more cluster thinning than selection 03 to achieve vine balance. Selection 04 tends to have heavier berries, more bunch rot, and poorer fruit composition than the others. Newer Italian imports, yet to be evaluated under California conditions, include Sangiovese FPS 07 (VCR 6), 08 (VCR 19), 09 (VCR 30), 10 (VCR 23), and 13 (VCR 102).

Production

Typically, production is 5 to 10 tons per acre. While 15 tons per acre can be achieved in warm, vigorous sites, it results in seriously delayed fruit maturation and poor fruit composition. Yields of 3 to 4 tons per acre are common in low-vigor sites. Crop load balance is essential to wine quality.

Harvest

Period: A mid- to late-season variety, harvested in mid-September to mid-October.

Method: The large, often well-exposed clusters make hand harvest easy. Canopy shaking results in easy to medium fruit removal with medium juicing. More whole clusters and cluster parts are removed than with trunk shaking. Trunk shaking results in easy to medium fruit removal with light juicing. Fruit is removed as single berries and some whole clusters.

leaves

Large; 3-lobed with large, triangular apical lobe; open U-shaped petiolar sinus; narrow lateral sinuses, inferior very shallow; short, sharp teeth; lower leaf surface mostly glabrous with scattered tufted hair.

shoot tips

Felty; young leaves green with yellow-bronze highlights.

Training and Pruning

Sangiovese is mostly trained to bilateral cordons and pruned to 12 to 16 two-node spurs. Higher node numbers may delay fruit maturation due to numerous, large clusters. Higher-vigor sites may utilize a quadrilateral cordon, usually requiring some cluster thinning. In Italy, growers believe it is important to have a large space (up to 10 inches) between spurs. Usually they prune to a single bud.

Trellising and Canopy Management

VSP is used in lower-vigor sites, while a single curtain, Smart-Dyson, GDC, or lyre system can be used in higher-vigor sites. Some shoot thinning and leaf removal may be needed, but the vine does not typically produce a dense canopy due to the moderately long internodes and minimal lateral shoot development. The clusters are sensitive to sunburn with excessive exposure.

Insect and Disease Problems

Summer bunch rot can be a problem in young vineyards with more compact clusters, especially in warmer climates. It is susceptible to Botrytis bunch rot with fall rains. Leafroll virus should always be avoided due to lower fruit anthocyanin content.

Other Cultural Characteristics

Crop load balance is the most important management concern due to Sangiovese's tendency to produce two or three clusters per shoot, with clusters averaging weights at $2/3$ to $1\frac{1}{3}$ pounds. Overcropping readily contributes to delayed fruit maturation and low fruit color, poor sugar/acid balance, and inferior wine aroma. Shoot thinning in the spring and/or cluster thinning at veraison are commonly practiced.

Winery Use

Styles range from rosé to full-bodied red wine, but most typically, Sangiovese is used for light- to medium-bodied Chianti-style wine. While 100 percent varietal wines are common, blends to add complexity and color are widely used. Blending varieties commonly used are Cabernet Sauvignon, Merlot, Cabernet franc, Zinfandel, or Ruby Cabernet, most often in percentages ranging from 10 to 20 percent.

—*L. Peter Christensen*

Sauvignon Blanc

Synonyms

In California, Fumé blanc and Savagnin Musque are used. In France, Sauvignon Jaune, Blanc Fumé, Fumé, Surin, and less commonly, Fié dans le Neuvillois, Punechon, Puinechou, and Gentin á Romorantin are used. Muskat Silvaner and Feigentraube are found in Germany.

Source

Sauvignon blanc has been grown for several centuries in Bordeaux and the Loire Valley. The exact origin is unknown.

Description

Clusters: small (occasionally medium on well-trained vines); cylindrical to globular; compact; short peduncles.

Berries: small; round to short oval; green-yellow when ripe; distinct "green pepper" flavor.

Leaves: medium; dark green; 3- to 5-lobed but overlapping lateral sinuses; U-shaped petiolar sinus; rounded teeth; leaf margin does not lay flat but is ruffled along the edge; sparse to moderate hair on lower leaf surface.

Shoot tips: felty white with rose margin; young leaves mostly green.

Growth and Soil Adaptability

Vines grow vigorously in many soil types in both cool and warm regions; it is generally advisable to avoid highly fertile and deep soils. Vine spacing should be a minimum of 6 feet. Shoots grow upright, which facilitates vertical-shoot-positioned trellises. Budbreak is after Chardonnay.

Rootstocks

Moderate- to low-vigor rootstocks are recommended to discourage additional vegetative growth. Water stress may occur late in the season with devigorating rootstocks if the irrigation strategy does not compensate for the high water demand caused by the large vine canopy.

Clones

Sauvignon blanc FPS 01 has a long history in California and for many years was the only registered selection available to nurseries. It came to the University of California in 1958 from Wente Vineyards with a history that indicated that it was imported in the 1800s from Chateau Yquem vineyard in France. This clone has performed well in California, but it is perhaps best known as the basis of the very successful New Zealand Sauvignon blanc industry.

Sauvignon blanc FPS 03 (Jackson), 22 (Oakville), 23 (Kendall-Jackson Winery), and 26 (Napa Valley) are all from California heritage sources. From Italy, Sauvignon blanc FPS 06 (ISV 5), 07 (ISV 2), 17 (ISV 1), and 24 (ISV 3) are now registered. In addition, the following French clones are now registered: Sauvignon blanc FPS 14 (French 316), 18 (French 317), 20 (French 242), 21 (French 378), and 25 (French 378).

clusters

Small (occasionally medium on well-trained vines); cylindrical to globular; compact; short peduncles.

berries

Small; round to short oval; green-yellow when ripe; distinct "green pepper" flavor.

Sauvignon blanc ENTAV-INRA® 376 is available as California certified stock. Sauvignon blanc ENTAV-INRA® 241, 297, 317, and 530 are available through the trademark program. Because most of these selections have only recently become available in California, very little comparative information is available about their performance.

In California, a selection of Sauvignon blanc with muscat flavor is often referred to as Sauvignon Musque: Recent DNA analysis has shown that at least two sources of California Sauvignon Musque are the same variety as Sauvignon blanc. One selection from France is available as Sauvignon blanc FPS 27. A California field selection of Sauvignon musque has been entered into the virus testing and registration program at UC Davis and should be available in the near future as a registered selection of Sauvignon blanc.

Production

This is a moderate-yielding variety averaging 5 to 7 tons per acre across most cool-climate sites. In warm climates, it will yield 8 to 12 tons per acre with quadrilateral cordon training.

Harvest

Period: An early season variety, harvested in September.

Method: The compact clusters are easy to hand harvest by field workers although the peduncle is short. Any type of machine harvester easily detaches intact berries from the vine, although juicing is moderate to high.

Training and Pruning

Head training with fruiting canes and renewal spurs is recommended due to low basal bud fertility; however, cordon training with spur pruning is common. Hedging is recommended no earlier than mid- to late July in order to avoid increasing berry size. Mechanical pre-pruning on cordon-trained vines facilitates the removal of canes that are held in the trellis wires by a large number of tendrils. Machine box pruning is successfully used in warm-climate districts.

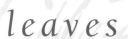

leaves

Medium; dark green; 3- to 5-lobed but overlapping lateral sinuses; U-shaped petiolar sinus; rounded teeth; leaf margin does not lay flat but is ruffled along the edge; sparse to moderate hair on lower leaf surface.

shoot tips

Felty white with rose margin; young leaves mostly green.

Trellising and Canopy Management

The fruiting wire height for a vertical-shoot-positioned system ought to be no more than 36 inches high because cane growth commonly will be no less than 48 inches long. In vigorous sites, a lyre or U-divided canopy is often used with at least a 3-foot split between canopies. Sunburn can occur on highly exposed fruit so a GDC is not recommended. Leaf removal around the clusters will enhance air movement and reduce the severity of bunch rot. However, this practice may also significantly affect fruit flavors.

Insect and Disease Problems

Latent viruses such as corky bark and stem pitting often become apparent when field-selected budwood is taken from Sauvignon blanc vineyards planted to AXR #1 rootstock and budded or grafted to other rootstocks. Older vineyards are often infected with leafroll virus that may delay ripening. The variety is susceptible to powdery mildew and Botrytis bunch rot; however, canopy management practices that increase air movement around the clusters may reduce disease severity. Cordon-trained vines are especially susceptible to Eutypa dieback.

Other Cultural Characteristics

The short internode length on canes results in a large number of tendrils on each shoot. Latent shoots are common, and shoot thinning is required on all training systems. Berry drop at set may occur if vegetative growth is excessive during that period. This is commonly associated with excess nitrogen from an untimely fertilizer application or cultivation of a cover crop.

Winery Use

As a varietal wine, Sauvignon blanc is dry or slightly sweet. It is commonly blended with Semillon. In a cooler climate it has a strong varietal character that is less pronounced in a warm area. If harvested late, or if noble rot occurs, it will produce a very sweet wine.

—*Rhonda J. Smith*

Semillon

Synonyms
Chevrier and Blanc doux are used in France; Greengrape in South Africa.

Source
Semillon is probably native to the Sauternes region of France, and it spread to other districts before the eighteenth century. It is grown throughout the world but its notoriety comes from the great white wines of Bordeaux. Semillon is typically blended with Sauvignon blanc to make both dry table wines and sweet dessert wines, including famous dessert wines from Sauternes. In California, Semillon is used in a similar fashion. It is a minor variety planted throughout the state.

Description
Clusters: medium; conical with shoulders, well-filled to compact, often winged; strong, long, woody peduncles.

Berries: medium to large; round to short oval; yellow to golden when ripe; characteristic date or fig-like taste. Some browning may occur on fully exposed fruit.

Leaves: medium: 3- to 5-lobed with relatively shallow inferior lateral sinuses; U-shaped petiolar sinus; small, sharp teeth; rough surface; margin rolled under; slight tufted hair on leaf underside.

Shoot tips: felty, with rose margin; young leaves yellow-green with bronze highlights.

Growth and Soil Adaptability
Semillon is a moderately vigorous variety with upright growth. It can be grown on a wide range of soil types. Its large clusters are prone to infection by *Botrytis cinerea*. Well-drained soils with moderate vigor potential are preferred. On high-vigor sites, dense canopies can lead to significant crop losses due to bunch rot.

Rootstocks
Semillon's moderate vigor allows for its use on any rootstock. Selection should be based on specifics of the site. When planted on poor soils, a vigorous stock such as St. George or 110R would be appropriate. On more fertile sites, Teleki 5C, SO4, 3309C, and 101-14 Mgt are all suitable.

Clones
Registered selections from California vineyards include Semillon FPS 02, 03, 04, 05, 06, and ENTAV-INRA® 173. Semillon ENTAV-INRA® 315, and 380 are available. No comparative studies of these clones have been made.

Production
Semillon produces moderately high yields, unless it is affected by Botrytis bunch rot. For late-harvest wines, fruit is allowed to remain on the vine until Botrytis develops.

clusters

Medium; conical with shoulders, well-filled to compact, often winged; strong, long, woody peduncles.

berries

Medium to large; round to short oval; yellow to golden when ripe; characteristic date or fig-like taste.

Harvest

Period: An early to midseason variety for table wine production. For dessert wines, Botrytis infections are desired, and harvest can occur several times over an extended period of time.

Method: Canopy shaking harvest is medium to hard. The soft pulp contributes to heavy juicing, especially with advanced maturity. Trunk shaking is easy to medium, with fruit removed as single berries and some cluster parts. Juicing is medium. Vertical-shoot-positioned systems improve harvestability.

Training and Pruning

Semillon should be cordon trained and spur pruned due to its large cluster size.

Trellising and Canopy Management

Semillon's upright growth lends itself to vertical-shoot-positioned systems. Only limited shoot positioning may be necessary. The use of split canopy systems should be considered only on sites with especially high-vigor potential. Leaf removal in the fruit zone can be used to improve the fruit zone microclimate and reduce the risk of Botrytis bunch rot.

Insect and Disease Problems

The large bunches are prone to Botrytis bunch rot.

leaves

Medium: 3- to 5-lobed with relatively shallow inferior lateral sinuses; U-shaped petiolar sinus; small, sharp teeth; rough surface; margin rolled under; slight tufted hair on leaf underside.

Felty, with rose margin; young leaves
yellow-green with bronze highlights.

Other Cultural Characteristics

The variety leafs out moderately late in spring
and may escape early frost damage.

Winery Use

Semillon is often blended into dry Sauvignon
blanc wines. Dessert wines are also produced,
especially in years with considerable Botrytis
bunch rot.

—*Edward Weber*

Syrah

Synonyms

Sirah, Syra, Schiras, Sirac, Syrac, and Petite Syras are used in France. It is called Shiraz in Australia and South Africa. In California, Syrah is not to be confused with "Petite Sirah" plantings, which are mostly of Durif (a hybrid of Peloursin and Syrah) but have included Peloursin and true Syrah.

Source

The true origin of Syrah has been shown through DNA testing to be a cross between two French varieties, Dureza and Mondeuse Blanche. Dureza is an obscure black variety and Mondeuse Blanche is a minor white variety, both of Rhône origin. Previous myths of origin included the Middle East (Shiraz, Persia); Roman importation into Gaul; Syracuse (Sicily); and Syrah Island, Greece. Syrah has been known in the Rhône Valley of France for many centuries where it has recently had a resurgence of popularity. Only 3,300 acres remained in 1958, but by the mid-1990s, plantings in southern France had increased to more than 86,000 acres. It is classified as recommended in the Rhône Valley, Provence, Languedoc, and southwest France; it is used in the production of AOC wines such as those of Hermitage, Cotes-du-Rhône, and Coteaux du Languedoc. The second largest plantings are in Australia where it is the leading red wine variety. Significant plantings also exist in South Africa and South America. Interest in the variety did not become widespread in California until the 1980s. It is now grown in a wide range of districts from the Central Valley and Sierra foothills to all but the coolest coastal districts.

Description

Clusters: medium; long cylindrical, loose to well-filled; very long peduncles causing the clusters to hang free from the canes.

Berries: small to medium; oval; blue-black; tend to shrivel when ripe.

Leaves: medium; mostly 3- to 5-lobed with reduced inferior lateral sinuses; U- to lyre-shaped petiolar sinus; short, sharp teeth; leaf surface occasionally bullate and puckered near petiole junction; tufted hair on lower leaf surface.

Shoot tips: felty with rose margin; young leaves yellowish with bronze highlights.

Growth and Soil Adaptability

Syrah is a very vigorous variety with a spreading growth habit and a tendency to produce long, trailing shoots. Growth can be excessive on deep, fertile soils and with high-vigor rootstocks. Recommended in-row spacing is 6 to 8 feet; suitable row middle spacing is 8 to 10 feet for bilateral cordon training and 11 to 12 feet for quadrilateral cordon training. Budbreak is fairly late.

Rootstocks

Freedom, Harmony, and Ramsey are suitable for nematode-prone sites if care is taken to match rootstock vigor to soil type, vine spacing, and trellis system. Drought-tolerant, phylloxera-resistant rootstocks should be

clusters

Medium; long cylindrical, loose to well-filled; very long peduncles causing the clusters to hang free from the canes.

berries

Small to medium; oval; blue-black; tend to shrivel when ripe.

used to minimize the fruit's tendency to shrivel during the latter stages of ripening. Good results have been experienced with 110R, 101-14 Mgt, and Kober 5BB. Suitable yields and vine growth have been seen with 3309C, SO4, and Schwarzmann.

Clones

The most widely distributed clonal material in California came from Australia as "Shiraz." It has been registered as virus tested in a series of numbered selections—Shiraz FPS 01 through 07—subclones that differ only on the length of their heat treatment. All of these selections have shown good viticultural and fruiting characteristics.

Other selections imported from France are adding to clonal diversity. Syrah FPS 04 (French 300), 05 (French 174), 06 (French 100), and 07 (French 877) are all available as generic clonal selections and registered stock. Syrah ENTAV-INRA® 525 is available as California certified stock. Syrah ENTAV-INRA® 99, 100, 174, 300, 388, 470, 471, and 877 are all available in the trademark program.

leaves

Medium; mostly 3- to 5-lobed with reduced inferior lateral sinuses; U- to lyre-shaped petiolar sinus; short, sharp teeth; leaf surface occasionally bullate and puckered near petiole junction; tufted hair on lower leaf surface.

Production

Syrah's yield potential is medium, usually 7 to 11 tons per acre, due to its relatively small berries, medium clusters, and low fruitfulness of basal buds.

Harvest

Period: A midseason variety, usually ripening in mid-September in the San Joaquin Valley, late September in the Sierra foothills, and late September to mid-October in the coastal districts.

Method: Hand harvest is moderately easy due to the long, green peduncles, but heavy growth can interfere, and the removal of numerous lightweight clusters is time-consuming. Canopy shaking results in easy to medium harvestability, mostly as single berries and some whole clusters and cluster parts, and with light to medium juicing. Trunk shaking results in easy to medium harvestability, mostly as single berries and some whole clusters, and with light juicing. Berry shriveling of ripe fruit reduces harvestability with any harvester.

Training and Pruning

Bilateral cordon training is most common with pruning to 8 to 20 spurs, depending on vine size. Higher-node numbers are sometimes used in highly vigorous vines. Syrah responds well to machine-hedge pruning with increased yields and little or no delay in fruit maturation. The method retains all nodes within the box configuration of the spur zone. Quadrilateral cordon

shoot tips

Felty with rose margin; young leaves yellowish with bronze highlights.

training with 20 to 30 spurs has been shown to be a favorable system to spread the vine canopy and fruiting zone in highly vigorous vineyards. Minimal pruning (no pruning except for canopy bottom) has been widely used in Australia, but lower berry skin anthocyanin content may occur.

Trellising and Canopy Management

Vigorously growing shoots are subject to wind breakage in the early spring. A foliar catch wire with bilateral cordons will reduce damage. High-vigor vineyards develop large canopies that respond favorably to horizontal quadrilateral systems such as GDC. It is not well suited to vertical-shoot-positioned systems in high-vigor sites due to the rapid growth of its long shoots. The trailing, downward growth from high cordons at 50 to 60 inches tends to moderate the rapid growth. VSP systems are suitable in cool districts or low-vigor sites.

Insect and Disease Problems

Syrah is moderately susceptible to powdery mildew, Phomopsis cane and leaf spot, and Pierce's disease. It has low susceptibility to sour rot except in highly vigorous, young vineyards where clusters may be very compact. Botrytis shoot blight and cluster blight from early spring rain can be a problem. It is not sensitive to Botrytis bunch rot from late season rains.

Other Cultural Characteristics

Berry shrivel is common at the latter stages of ripening, often beginning at 21 to 22° Brix. This contributes to increased soluble solids but declining harvest weight and increased difficulty of machine harvest. Therefore, harvest must be timely as the grapes reach optimum maturity. Phloem flow of water and solutes into berries is apparently impeded once they reach maximum weight. Thus, berry weight loss and soluble solids concentration are the result of berry transpiration.

Syrah is susceptible to lime-induced chlorosis and "spring fever," a springtime nitrogen metabolism disorder that occurs during fluctuating warm and cool periods during rapid shoot growth before bloom. Basal leaves will fade in green color from the edges to between the primary and secondary veins. This is often accompanied by a border of red to purplish pigmentation and marginal leaf burn. The symptoms are associated with elevated levels of putrescene, a polyamine, although they are sometimes mistakenly diagnosed as potassium deficiency. It is a seasonal, non-debilitating phenomenon; there are no known preventative measures.

Winery Use

A versatile variety, Syrah is well adapted to a wide range of viticultural temperature regions, winery uses, and wine styles. Used to produce varietal table wines of distinct character in the cooler districts, it also has demonstrated high potential for red table wine production in the warmer districts, including the San Joaquin Valley. It has good blending qualities for deep color and not overly tannic, fruity aromas, producing popular blends such as Cabernet Sauvignon-Shiraz, an Australian conception, as well as traditional Rhône blends. Dessert wine potential is high, with Australian Port-type wines as good precedents.

—*L. Peter Christensen*

Tempranillo

Synonyms

In Spain, it is known as Tempranillo de la Rioja, Tinto de la Rioja, Tinto del Pais, Grenache de Logrono, Jacivera, Tinto de Toro, and Tinto Madrid. In the La Mancha region of Spain, it is called Cencibel; in Ribero del Duero, Tinto Fino Ull de Llebre, and in Catalonia, Ojo de Liebre. In Portugal, it is called Tinta Roriz and Aragonez. Previously it was known as Valdepeñas in California.

Source

The variety is most likely a selection from northern Spain, but some believe that Tempranillo originated in southern France as a natural hybrid of Cabernet franc and Pinot noir.

Description

Clusters: medium to large; cylindrical to long conical, compact; medium-length peduncles.

Berries: medium; round to pear-shaped with a flat apex; deep blue-black.

Leaves: large; deeply 5-lobed with overlapping lateral lobes, lyre-shaped petiolar sinus; large, sharp teeth; moderate to dense tufted hair.

Shoot tips: felty with rose margins; young leaves yellow-green with bronze-red patches.

Growth and Soil Adaptability

Vines have moderately high vigor on deep, fine sandy loam to clay loam soils. Budbreak tends to be similar to Zinfandel but ripens earlier. The shoots are semi-erect and the canopy is more open than that of Zinfandel. Tempranillo seems to grow well where Zinfandel does well. The vine appearance is similar to Zinfandel's, and the two may be found mixed into old vineyard plantings. Vine spacing should be about 8 to 10 feet down the vine row for vertical-shoot positioning or standard bilateral cordons. Spacing down the row of less than 8 feet may be appropriate on low-vigor rootstocks or in sites of poor soil. For horizontally divided quadrilateral vines, spacing should be 6 to 7 feet. Moderate-vigor rootstocks should be used for most sites.

Rootstocks

Moderate- to low-vigor rootstocks are better choices, especially if close spacing is desired. Many older established blocks are mostly own-rooted vines. Freedom may be acceptable where nematodes are a concern; better overall choices include 101-14 Mgt, Kober 5BB, 110R, or 1103P. For close spacing, or where less vigor is desired, 3309C, Schwarzmann, or 1616C may be needed.

clusters

Medium to large; cylindrical to long conical, compact; medium-length peduncles.

berries

Medium; round to pear-shaped with a flat apex; deep blue-black.

Clones

Tempranillo FPS 02 and 03, from Spanish sources, are available from UC Davis. Under another synonym for Tempranillo, registered Valdepeñas FPS 03 was established from the University's Jackson Field Station collection. In addition, Tinta Roriz 01 (also a synonym for Tempranillo) is registered at FPS; Harold Olmo imported this selection from Portugal in 1984. No viticultural or enological records exist to indicate distinctive differences in growth, production, or resultant wines.

Production

Tempranillo vines are productive to very productive, capable of bearing medium to large crops of 8 to 12 tons per acre. High yields may sacrifice color intensity and fruit flavors, significantly reduce acid level, and increase pH while delaying harvest.

Harvest

Period: An early season variety, harvested in late August to mid-September.

Method: Hand harvest is facilitated by large clusters, which are often free-hanging and accessible, with an easily cut cluster stem. Machine harvest by trunk shaker is intermediate in difficulty. The potential for juicing of berries at harvest is light to medium, but not severe as berry skins are relatively tough. Harvest by pivotal striker is also intermediate, but newer trunk shaker or canopy shaker (bow-rod) heads may reduce the possibility of juice losses.

Training and Pruning

Tempranillo is well suited to spur pruning and bilateral cordon or quadrilateral training. It is also well suited to head training (vertical cordon). A spur count of 14 to 20 two-node spurs is acceptable, depending on rootstock, soil depth, and soil texture. Cane pruning is not suggested but may work with cluster thinning.

leaves

Large; deeply 5-lobed with overlapping lateral lobes, lyre-shaped petiolar sinus; large, sharp teeth; moderate to dense tufted hair.

Felty with rose margins; young leaves
yellow-green with bronze-red patches.

Trellising and Canopy Management

Vertical-shoot-positioned systems are recommended. High-vigor sites may benefit from a divided canopy system such as a quadrilateral cordon or GDC type trellis, but cross-arm foliage wire may be required to avoid excessive fruit exposure in hotter regions.

Insect and Disease Problems

The variety is moderately sensitive to powdery mildew and downy mildew but very resistant to any bunch rot, either Botrytis or sour bunch rot. It is very susceptible to Eutypa dieback disease.

Other Cultural Characteristics

At budbreak Tempranillo is above average in its tolerance to cold spring temperatures. Berries can sometimes set irregularly, especially on sandy, zinc-deficient soils. Berry skins are relatively tough. Fruit holds well on the vine during the late stages of maturity. Acid levels can be low and pH marginally high in very warm years or hotter regions. Fruit color is better and develops more uniformly compared to Zinfandel.

Winery Use

Tempranillo produces good- to excellent-quality wines with good color under optimum conditions, especially at lower crop levels. The wine can have distinctive varietal character. Its uses range from good blending varietal to high-quality varietal table- or port-wine blends. Future interest in Tempranillo as a premium wine should increase for all areas, although lack of marketing and consumer awareness may limit its overall importance.

—Paul S. Verdegaal

Valdiguié

Synonyms

In California, Valdiguié was misnamed and known as Napa Gamay (or simply Gamay) until 1980 when it was properly identified by Pierre Galet. Its identity has since been verified by DNA fingerprinting. It is still listed as Gamay in the California official acreage reports, but the above names are no longer allowed on wine labels. Some California wineries are now labeling the wine as Valdiguié, the official name in France. However, Valdiguié fruit still may be used in wines labeled Gamay Beaujolais until 2007. Synonyms in France include Valdiguer, Cahors, Gros Auxerrois, Jean-Pierrou at Sauzet, Quercy, and Noir de Chartres.

Source

Valdiguié's French origin is unclear due to conflicting claims of its first cultivation. It was first commercially propagated in 1874. Plantings became fairly widespread through-out southwestern France, largely because of its high yield and powdery mildew tolerance. It is still grown in Languedoc and Provence, but as a minor variety. In California the variety gained popularity during Prohibition because of its high productivity, with acreage scattered throughout Northern California after Repeal. Areas of later expansion included the Central Coast and Central Valley, as well as the North Coast. Following the wine grape planting boom of the 1970s, total plantings peaked at 6,118 total acres in 1977. Total acreage has since declined to below 1,000 acres.

Description

Clusters: medium to large; long conical, well-filled to compact; medium-size peduncles.

Berries: medium to large; short oval; dark blue-black with white bloom.

Leaves: medium; mostly entire with shallow lateral sinuses; closed U-shaped petiolar sinus; short, sharp teeth; lower leaf surface with sparse, lightly tufted hair.

Shoot tips: felty white; young leaves yellow-ish and downy.

Vines are moderately weak in appearance, becoming gray-green to yellowed near harvest.

Growth and Soil Adaptability

Vines are moderately vigorous to vigorous, unless depressed with heavy cropping. With moderate growth, shoots will remain upright and the canopy will spread with non-shoot-positioned vines. Vigorous vines will grow upright during the early season, but the shoots will trail by late season. Valdiguié is suited to a fairly wide range of soil types. It is adapted to the warmer districts of the North and Central Coast. The vines leaf out fairly late, tending to escape early spring frosts.

clusters
Medium to large; long conical, well-filled to compact; medium-size peduncles.

berries
Medium to large; short oval; dark blue-black with white bloom.

Rootstocks

Vine vigor can be excessive with vigorous root-stocks. Therefore, moderately vigorous rootstocks such as SO4, Teleki 5C, 420A, 3309C, and 1616C are preferred unless soil fertility is limited.

Clones

Registered clones in California are presently listed as Napa Gamay FPS 01, 02, and 03. FPS Napa Gamay selection 01 is from a California vineyard. It was heat treated to produce two additional registered selections: Napa Gamay 02 (heat treated 63 days) and Napa Gamay 03 (heat treated 102 days). None have been performance tested.

A selection is in the registration process at FPS from a Napa Valley heritage vineyard. There are no Valdiguié clones listed in the French official clonal registries.

Production

Vines produce 5 to 8 tons per acre. Its abundant and large clusters can easily contribute to over-cropping, especially with young vines.

Harvest

Period: A mid- to late-season variety, harvest occurs in mid-October to early November in the cooler districts (North and Central Coast) and in early to mid-October in warmer, interior districts.

Method: Hand harvest is easy due the grape's large clusters. The short and hard peduncles often require clippers for removal. Machine harvest with canopy shakers is medium, with fruit removed mostly as single berries. Juicing is medium. Trunk shaking is easy to medium, with fruit removed as cluster parts as well as single berries; juicing is light to medium.

leaves

Medium; mostly entire with shallow lateral sinuses; closed U-shaped petiolar sinus; short, sharp teeth; lower leaf surface with sparse, lightly tufted hair.

shoot tips

Felty white; young leaves yellowish and downy.

Training and Pruning

Some older vineyards are head trained with spur pruning. Newer vineyards are bilateral cordon trained with 8 to 14 spurs per vine, depending on vine size. Cane pruning and quadrilateral cordon training are not recommended due to the vine's large clusters and tendency to overcrop.

Trellising and Canopy Management

The upright growth makes the variety well adapted to vertical-shoot-positioned systems. However, the vigorous growth and large leaves may require shoot and leaf removal and canopy trimming.

Insect and Disease Problems

Older vineyards are commonly infected with leafroll virus. Only certified planting stock should be used for new plantings. The variety shows a modest degree of tolerance to powdery mildew.

Other Cultural Characteristics

Crop thinning may be necessary to avoid over-cropping, especially in cooler districts where fruit ripening is late.

Winery Use

Valdiguié is used for the production of fruity red or rosé table wines under a varietal label. Fruit anthocyanin content is adequate for light- to medium-bodied red wines in coastal districts but may be insufficient in the Central Valley. "Nouveau" wines of Valdiguié have used carbonic maceration, a fermentation method using uncrushed grapes in closed fermenters to produce light-bodied wines to be sold when young.

—*L. Peter Christensen*

Viognier

Synonyms

In France, Viognier is called Bergeron, Barbin, Rebelot, Greffou, Picotin Blanc, Vionnier, Petiti Vionnier, Viogné, Galopine, and Vugava bijela.

Source

Viognier is from Southern France, in the Rhône Valley districts of Condrieu and Château-Grillet. Some people believe that the vine originally was brought to France by the Roman Emperor Probus from the Dalmatia region where it is now cultivated under the name Vugava bijela. The areas now planted to Viognier are increasing worldwide.

Description

Clusters: medium, long-cylindrical with broad shoulders, well-filled to compact; medium peduncles.

Berries: small; round to short oval; yellow and amber when ripe; distinct aromatic flavor when fully ripe.

Leaves: medium; mostly 3-lobed with wide U-shaped petiolar sinus and reduced inferior lateral sinuses; medium-length, sharp teeth; slightly bullate surface; light- to moderately tufted hair on lower surface.

Shoot tips: downy to felty white with rose margin; green, young leaves with slight bronze highlights.

Growth and Soil Adaptability

Viognier is a low-moderate vigorous vine, but it can be productive under vigorous conditions. In France, it is planted on steep, shallow granitic soils. In the North Coast of California, it is usually planted on deep, alluvial soils. Viognier is not a good choice for shallow, dry soils. The canopy is somewhat open, with slender, pendant canes that need support. Since the vines are only moderately vigorous, close spacings are suitable: in moderate soils plant 4 to 6 feet in row, and in deep soils plant 6 to 8 feet in row.

Rootstocks

In France, Viognier is most often propagated on 110R rootstock. In the United States, Teleki 5C, SO4, 3309C, and 101-14 Mgt rootstocks are used in coastal regions. In inland valleys and on fertile soils, Freedom, Kober 5BB, SO4, Teleki 5C, 110R, and 101-14 Mgt are all good choices depending on spacing, trellising, and soils.

Clones

There is a single French clone certified, ENTAV-INRA® 642. In addition, Viognier FPS 01 was obtained from France during the 1970s, and it is believed to be clonally distinct. Both are commercially available in California. Non-registered field selections have often had severe virus problems. Recently, three Roussanne selections donated to FPS were identified as Viognier (FPS 02, 03, and 04). All three are expected to advance to registered status by 2005.

clusters

Medium, long-cylindrical with broad shoulders, well-filled to compact; medium peduncles.

berries

Small; round to short oval; yellow and amber when ripe; distinct aromatic flavor when fully ripe.

Production

Viognier is a moderate- to good-yielding cultivar. In France, yields are among the lowest of any wine grape, averaging about 1.3 tons per acre. In California, yields of 1.5 to 4 tons per acre have been reported in coastal regions while 5 to 8 tons are possible in the northern interior valley and foothills. The most productive coastal vineyards might reach a yield of 5 tons per acre in a good season.

Harvest

Period: An early variety, ripening around the same time as Chardonnay; in the North Coast, this is early to mid-September. A late September harvest may be needed to satisfy many winemakers' demand for 25 to 26° Brix maturity.

Method: The clusters are small, making hand harvesting slow. Canopy shaking is easy to medium, with whole berries and many cluster parts removed, and light to medium juicing.

Training and Pruning

In France, the Guyot system is used. In the United States, both cane and spur systems are used, canes particularly in cooler regions. A commonly used system is bilateral cordons with closely spaced spurs (4 to 6 inches), each spur with two nodes. Cane pruning is less common. Two canes up to 30 inches long with 8 to 12 buds each are typical, along with two renewal spurs pruned to two nodes. Viognier is well adapted to a vertical-shoot-positioned system but does well on standard bilateral with foliage catch wiring with a "T" or "Y" top.

Trellising and Canopy Management

In California, vertical-shoot-positioned systems work well. Some growers have tried divided-canopy trellises on deeper soils, but the vines usually don't appear to be vigorous enough for this system except on very fertile soils or with the most vigorous rootstocks such as 140Ru, 110R, 1103P, or Freedom.

leaves

Medium; mostly 3-lobed with wide U-shaped petiolar sinus and reduced inferior lateral sinuses; medium-length, sharp teeth; slightly bullate surface; light- to moderately tufted hair on lower surface.

shoot tips
*Downy to felty white with rose margin;
green, young leaves with slight bronze
highlights.*

Insect and Disease Problems

Viognier is moderately sensitive to powdery mildew. The French consider it resistant to Botrytis bunch rot; California experience seems to confirm this.

Other Cultural Characteristics

Viognier can have clusters with both small and large berries (known as "hens and chickens"). To achieve maximum flavors, this variety should be planted in warm areas, such as high Winkler Region II, Region III, and low Region IV, and thoroughly ripened to above 24° Brix.

Winery Use

Viognier produces very fragrant (floral and fruity aromas) wines with good acidity, tannin structure, and relatively high alcohol. The best examples are produced fairly simply and cleanly, often going from stainless steel fermentation to the bottle. When properly made, these wines age well in the bottle. Techniques used for Chardonnay (barrel fermentations, malolactic fermentations, aging on lees) may mask the unique flavor and fragrance of this variety, probably due to a high terpene content of the fruit. However, some vintners prefer to use oak fermentation for an alternative and desirable wine style. Viognier is sometimes blended with Syrah (up to 20 percent) to give the resulting wine more fragrance and elegance. Some French Château Viogniers are considered to be among the best and most expensive white wines in the world.

—*Glenn McGourty*

Zinfandel

Synonyms
None

Source

Zinfandel is only grown under this name in California. As a result, historians have long debated the appearance of this variety in the state. Some believe Zinfandel was first imported from Hungary in 1852; others point to evidence that New England nurseries had cultured it as a table grape and introduced the variety in California during that decade. By the mid-1860s, wines made from Zinfandel grapes were seen as an improvement to those made from the popular Mission variety, and plantings of Zinfandel vines increased.

It was also speculated that Zinfandel came from southern Italy, as a similar variety, Primitivo, was found to have identical DNA. Most recently, DNA analysis has shown Zinfandel to be identical to a very obscure variety found in Croatia called Crljenak Kasteljanski, which translates to "the red grape of Kastela." Crljenak probably originated in the Balkans on the Dalmatian coast. This is supported by numerous genetic relatives found in the area through DNA testing. One such relative, Plavac Mali, has been shown to be a cross between Zinfandel and a rare Croatian variety, Dobricic. Exactly how Zinfandel came to California is still unknown.

Description

Clusters: medium to large; cylindrical to long conical, often winged, sometimes with double wings, compact; short to medium-length peduncles; often with a wide range of ripe and under-ripe berries.

Berries: medium to large; round to oblate; deep blue-black; prominent rust-colored stylar scar.

Leaves: medium to large; deeply 5-lobed, often overlapping; lyre-shaped petiolar sinus; long, jagged teeth; dense hair on lower leaf surface.

Shoot tips: downy to felty; young leaves bronze-red.

Growth and Soil Adaptability

Zinfandel is currently grown in many of the wine-growing regions of California in a wide range of soil types and climates. In fertile sites it must be managed carefully to avoid overcropping due to its naturally high fruitfulness. It is best suited for moderate- to low-fertility, well-drained soils where it is considered moderately vigorous if provided with supplemental irrigation. Vine spacing should be no less than 5 feet and should be increased as soil fertility increases.

clusters

Medium to large; cylindrical to long conical, often winged, sometimes with double wings, compact; short to medium-length peduncles; often with a wide range of ripe and under-ripe berries.

berries

Medium to large; round to oblate; deep blue-black; prominent rust-colored stylar scar.

Rootstocks

Historically this variety was grown on its own roots or grafted to *Vitis rupestris* St. George. This practice continues in hillside plantings or on sites with limited irrigation water. A drought-tolerant rootstock, 110R, is also used in these sites. It can be successfully grown on other rootstocks as long as they do not impart increased fruitfulness to the scion since Zinfandel tends to set large clusters. Many older selections of Zinfandel grown on St. George are not certified, and latent viruses may cause disease when grafted to other rootstocks.

Clones

Zinfandel FPS 01, 02, 03, and 06 have been registered for many years. They have a poor reputation among some winemakers due to large clusters and berries, poor fruit color, and lack of varietal character. In addition, all of these selections are susceptible to bunch rot. In warm climates, research has shown that there are very few differences in growth and yield parameters among the registered selections of Zinfandel and only subtle differences in wine quality. Primitivo, a variety grown in southern Italy for 150–250 years, and Zinfandel are now considered to be clones of the same variety. When compared to Zinfandel, Primitivo has increased cluster numbers yet reduced yields due to fewer and smaller berries. Primitivo fruit ripens before Zinfandel and is much less prone to bunch rot. Primitivo 03, 05, and 06 are all commercially available as certified selections.

A heritage Zinfandel clonal program is underway at the University of California. Researchers are collecting budwood from vineyards that are at least 60 years old and preserving it in a secure collection. Selections are introduced into a long-term clonal evaluation program. This project is coordinated with simultaneous virus testing and registration of these selections. Preliminary results indicate a range of performance in yield parameters across the evaluated selections and the current registered Zinfandel selections are clustered around the mean of this range. The lack of popularity of the older registered selections may also be due to perceived problems in site and vine management practices.

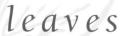

leaves

Medium to large; deeply 5-lobed, often overlapping; lyre-shaped petiolar sinus; long, jagged teeth; dense hair on lower leaf surface.

shoot tips

Downy to felty; young leaves bronze-red.

Production

Production is extremely variable throughout California and is dependent upon climate, soil fertility, crop level management practices, and irrigation. In addition, yields in vineyards that are harvested for white wine production will tend to be higher than yields of vineyards farmed for red wine production since lower sugar levels are acceptable in the former. In the Sacramento Valley and the Lodi area where supplemental irrigation is ample, grape yields for red wine production can range from 6 to 8 tons per acre in trellised vineyards with vines trained to bilateral cordons. Head-trained, spur-pruned vineyards will yield 3 to 6 tons per acre. In coastal or foothill Zinfandel vineyards farmed for red wine production, cluster thinning is common to maximize crop uniformity for color and ripeness. Yields for trellised vineyards could range from 5 to 6 tons per acre, with 4 tons not being uncommon. In these regions, typical yields for head-trained, spur-pruned vineyards would be 3 to 5 tons.

Harvest

Period: A mid- to late-season variety, although the harvest period depends on crop load and whether the fruit is going for red or white wine. In the North Coast and Sierra foothills, grapes for red wine production are harvested midseason (September) while harvest of vines producing white Zinfandel will begin in early to mid-August.

Method: Harvesting is most commonly by hand. The clusters are large and easy to pick unless bunch rot is severe and then infected fruit must be cut out of the bunches. Machine harvesting is more common with new vineyards established on trellises. Zinfandel tends to juice moderately easily, but levels are still acceptable for machine harvesting.

Training and Pruning

Vines are spur pruned and either cordon or head trained. Zinfandel tends to overcrop easily, and if the fruit is not thinned, it will ripen with difficulty or not at all. As a result, many growers note that the fewer spurs there are, the less potential there is to accidentally allow too many clusters to remain on the vine. A head-trained, spur-pruned vine by default cannot have as many spurs as a bilateral cordon trained vine.

Trellising and Canopy Management

Most vineyards have trellises that support a fruiting wire for the cordons and either two pairs of moveable wires to position the shoots vertically or a cross arm with a foliar support wire at each end. Cordons are 38 to 42 inches high. Shoot thinning is common as a means of crop load adjustment and encouraging uniformity of development. Depending on the year, clusters may be thinned prior to or at veraison, and some growers consistently drop clusters after veraison. Leaves are removed from the fruiting zone to minimize a microclimate conducive to bunch rot and improve fruit quality; however, excessive sun exposure will sunburn fruit.

Insect and Disease Problems

Zinfandel's compact clusters are susceptible to physical damage, insect damage, or disease. Bunch rot is hard to avoid. Its severity may be reduced with leaf removal and shoot positioning. Insects that feed or lay eggs in the clusters such as omnivorous leafroller and orange tortrix will increase the incidence of bunch rot. Willamette mite infestations are common in the north half of the state and the coast, while Pacific mite is a problem in the southern San Joaquin Valley. Older vineyards are often infected with leafroll virus that may delay ripening. In some instances, this delay may cause the crop to be consistently sold for white Zinfandel wine production. Zinfandel is moderately resistant to powdery mildew and Eutypa dieback.

Other Cultural Characteristics

Berry size is affected by water availability and irrigation strategy. If water is applied incorrectly, berry size will increase and bunch rot will nearly always occur. Uneven ripening is common in any climate, and the clusters contain immature pink berries as well as overripe, slightly wrinkled berries. In very warm climates a significant portion of the cluster may contain shriveled berries or raisins.

Winery Use

Winemakers produce a single varietal Zinfandel wine or may prefer to add small amounts of other varieties, commonly found in old, mixed plantings, to enhance complexity. In cooler areas, the fruit will produce wines that have berry fruit and spice flavors while in warm areas varietal character is less obvious. Zinfandel can reach high sugar levels and, as a result, can produce a high-alcohol table wine or port-style dessert wine.

—*Rhonda J. Smith*

Minor

Wine Grape Varieties

in California

Minor Wine Grape Varieties
in California—*Black*

	Aglianico	Alicante Bouschet
Origin	South-central Italy	A French hybrid grape (Petit Bouschet × Grenache) by Henri Bouschet in 1886
Use	For high-quality, red table wines	To intensify red color; mainly shipped for home winemaking
Clusters	Medium; conical, compact; medium-length peduncles	Medium; broad conical, well-filled; long peduncles
Berries	Medium; short oval; blue-black with relatively dense gray-blue bloom	Medium to large; round berries; red juice; berries much larger than Rubired
Leaves	Small to medium; mostly 3-lobed; closed U-shaped petiolar sinus; shallow inferior lateral sinuses; teeth relatively common in superior lateral sinuses; medium-long, sharp teeth; moderately dense tufted hair	Medium; mostly entire with reduced lateral sinuses; closed U-shaped petiolar sinus; leaf margins roll under
Shoot tips	Downy, green; young leaves green with pink highlights	Felty; reddish young leaves and shoots
Growth	Semi-upright habit; medium vigor	Semi-erect to spreading habit. Medium vigor with moderately open canopy
Adaptability	Warm regions. Color may be deficient in hot regions; heat summation for ripening and balance may be insufficient in cool regions	To warm to hot regions
Yield	5 to 9 t/a	8 to 11 t/a
Harvest period	Late Sept. to late Oct.	Late Sept. to early Oct.
Harvest method	No reported experience	Medium to hard harvestability (canopy shaker); medium to heavy juicing
Culture	Close to medium spacing; trained to bilateral cordons and spur pruned. VSP and overhead (pergola) systems used in Italy	Medium spacing; trained to bilateral cordons and spur pruned. Vineyards often head-trained for fresh shipment.
Other characteristics	Sensitive to drought but prone to bunch rot with excessive moisture. Wide genetic variability of vine vigor and fruit size in Italian vineyards where Taurasi-type clones are preferred. Leafroll virus-free wood sources are essential.	Susceptible to vine dieback and delayed foliation from overcropping. Older vineyards are infected with leafroll or yellow vein virus. Susceptible to spider mites and bunch (sour) rot.

Carnelian	Centurion
Cross of (Carignane × Cabernet Sauvignon) × Grenache by H.P. Olmo. UC patent 3625 in 1973	Cross of (Carignane × Cabernet Sauvignon) × Grenache by H.P. Olmo. UC patent 3870 in 1975
For fruity, light to full-bodied red table wines	For light- to medium-bodied red or rosé table wines
Medium to large; broad conical, well-filled; medium-thick peduncles; first and second cluster equal in size and often intertwine	Medium to large; broad conical, well-filled; long peduncles; very fruitful
Small to medium; round; dark purple-black	Small; round; purple-black
Medium to large; 3-lobed with overlapping lateral sinuses to give mostly entire appearance; wide U-shaped petiolar sinus; leaf surface glabrous and shiny	Medium; 3-lobed to mostly entire (reduced lateral sinuses); U-shaped petiolar sinus; medium-length, rounded teeth; shiny, upper leaf surface; sparse tufted hair on lower leaf surface
Felty, light-cream colored with pink margin; young leaves light yellow-green and very glossy	Downy; young leaves green with pink, not bronzed, borders
Erect; similar to Grenache; high vigor with moderately dense canopy	Semi-erect to erect growth habit; medium-high vigor with a moderately dense canopy
Warm to hot regions	Warm to hot regions
10 to 13 t/a	10 to 14 t/a
Early to late Sept.	Early to late Sept.
Hard harvestability, mostly single berries; medium juicing	Medium (canopy shaker) to medium-hard (trunk shaker) harvestability; light to medium juicing
Medium to wide spacing; trained to bilateral cordons and spur pruned	Medium to wide spacing; trained to bilateral cordons and spur pruned
Young vines tend to overcrop; delayed foliation and dieback may result. Fruit may attain high sugar levels.	Young vines tend to overcrop; crop control by early cluster thinning is often necessary. Fruit may attain high sugar levels (25° Brix).

Minor Wine Grape Varieties
in California—*Black* (continued)

	Cinsaut	Dolcetto
Origin	Southern France	Piedmont region, Italy
Use	For softness in blends of quality red table wines; formerly known as Black Malvoisie in California	For light-bodied, red table wines
Clusters	Medium to large; well-filled, long conical; long peduncles	Medium; long conical, loose to well-filled; commonly very loose at cluster end
Berries	Large; long oval, purple-black	Small to medium; round; blue-black and lighter red when exposed or in light
Leaves	Medium; deeply 5-lobed with U-shaped petiolar sinus; lateral sinuses have V-notch at base and occasional teeth in superior lateral sinuses; long, sharp teeth; sparse, cobwebby tufted hairs on lower leaf surface	Large; 5-lobed with wide U-shaped petiolar sinus; medium-length, sharp teeth; smooth upper surface and glabrous on lower leaf surface; veins and petioles dark red-violet
Shoot tips	Felty white with pink-red margin; young leaves deep green with bronze-red highlights	Wooly and white with red margin; red-violet young leaves
Growth	Spreading habit; medium vigor with moderately open canopy	Spreading; low to medium vigor, often with sparse, open canopy
Adaptability	Warm to hot climates, but typically with light-colored red wines; cooler climates contribute to acceptable red wine color	Cool to warm regions; unfavorable in hot regions
Yield	6 to 12 t/a	5 to 8 t/a
Harvest period	Early Sept. to early Oct.	Late Sept. to mid Oct.
Harvest method	Easy harvestability; light to medium juicing	Easy harvestability; medium juicing
Culture	Medium vine spacing; trained to bilateral cordons and spur pruned. Foliar support trellis may be needed due to thin shoots and cluster exposure. High soil fertility and water status contribute to poor fruit color.	Close to medium spacing; trained to bilateral cordons and spur pruned; vertical trellising appropriate for most sites
Other characteristics	Large berries contribute to low wine color and tannin content.	High cluster numbers easily contribute to over-cropping; cluster thinning often warranted; susceptible to berry shrivel and berry drop near harvest.

Freisa	Mission
Piedmont region, Italy	Spain. Introduced as the first V. *vinfera* variety in California by Mission Fathers (1778) and the principal wine grape until about 1870. Similar to Pais of Chile and Criollo of Argentina.
For dry, light-red wines to dry, sparkling, or semi-sweet sparkling wines, sometimes of low alcohol	Mostly for dessert wines such as sherry- or port-type
Small to medium; long conical to cylindrical, often winged to double, well-filled to compact; long peduncles	Large; long conical, loose to well-filled; medium to long peduncles
Medium; oval; purple-black	Medium; round to oblate; light red-purple
Medium; 3-lobed to almost entire; very wide U-shaped petiolar sinus; shallow lateral sinuses; short, sharp teeth; glabrous lower leaf surface	Large; 5-lobed with open U-shaped petiolar sinus; short, sharp teeth; leaf surface glabrous and smooth; short, tufted hair on lower leaf surface
Cobwebby and yellow-green with slight copper color; green-yellow young leaves with slight bronze-red highlights	Felty without rose margin; young leaves green with slight bronze-red coloration
Trailing habit; medium vigor with long internodes and open canopy; similar to Nebbiolo	Semi-upright with long shoots. High vigor; dense canopy.
Warm regions	Warm and hot regions
6 to 10 t/a	8 to 13 t/a
Mid Sept. to early Oct.	Late Sept. to late Oct.
No California experience	Medium (canopy shaker) to easy (trunk shaker) harvestability; trunk shaker preferred; medium to light juicing
Medium spacing; bilateral cordons with spur pruning. VSP trellising and cane pruning used in Italy; leaf removal usually not required.	Medium to wide spacing; trained to bilateral cordons and spur pruned
High productivity may require cluster thinning. Musts are typically high in acidity and tannins and moderate to good in anthocyanins.	Susceptible to Pierce's disease and crown gall. Fruit holds well on vine after ripening. Table wines lack color, acidity, and aroma.

Minor Wine Grape Varieties
in California—*Black* (continued)

	Montepulciano	Petit Verdot
Origin	Central Italy	Medoc, southwest France
Use	For light- to full-bodied red table wines and for blending	Mostly blended with red Bordeaux varieties for color, tannin, and complexity.
Clusters	Small; conical, well-filled to compact; small to medium peduncles	Small to medium; cylindrical often winged to double; well-filled to compact; short- to medium-length peduncles
Berries	Small; round; purple-black	Small; round; purple-black
Leaves	Medium to large; very deeply 5-lobed, with U-shaped petiolar sinus; long, sharp teeth; moderately dense hair on lower leaf surface	Medium; mostly 3-lobed with reduced inferior lateral sinuses; U-shaped petiolar sinus often with teeth along edge; short teeth; dense hair on lower leaf surface
Shoot tips	Felty and gray-white; young leaves green with bronze-red highlights	Felty; young leaves yellow-green with slight bronze-red highlights
Growth	Trailing; high vigor	Semi-upright to trailing; medium-high vigor
Adaptability	Warm regions	Cool and warm regions
Yield	8 to 10 t/a	5 to 7 t/a
Harvest period	Mid Oct.	Late Sept. to late Oct.
Harvest method	No reported experience	No reported experience
Culture	Medium spacing; bilateral cordon training with spur pruning. Overhead (pergola) and VSP trellising are used in Italy.	Medium spacing; bilateral or quadrilateral cordon training with spur pruning. GDC is suitable in high-vigor sites due to somewhat dense canopy of trailing growth.
Other characteristics	Higher acidity and phenolic content than Sangiovese; known for versatility in wine styles; very fruitful; overcropping can greatly decrease wine quality.	Sensitive to powdery mildew; low bunch rot potential. FPS selection 02 is more productive than selection 01.

Tannat	Tinta Madeira
Southwestern France, Pyrenean area	Portugal, possibly the island of Madeira or Alto Duro Valley
For quality red table wines, mostly in blends for color, acidity, and tannin	For premium port-style wines
Medium; cylindrical often winged, well-filled to compact; short- to medium-length peduncles	Medium; conical, well-filled to compact; medium-length peduncles
Small; round; purple-black; very small berries for size of cluster	Medium; oval shaped; deep purple with dense white bloom
Medium; mostly 5-lobed with closed U-shaped petiolar sinus; small, short teeth; puckered bullate leaf surface; moderately dense, tufted hair on lower leaf surface	Medium; mostly 3-lobed with reduced inferior lateral sinuses; closed U-shaped petiolar sinus; short, sharp teeth; moderately dense tufted hair on lower leaf surface
Felty white with rose margin; young leaves reddish with bronze spots	Downy white-green; young leaves yellow-green with very slight bronze highlights
Semi-upright to trailing. High vigor, but less than Cabernet Sauvignon	Trailing; medium vigor with moderately open canopy
Cool to warm regions	Warm to hot regions
6 to 13 t/a	6 to 9 t/a
Late Sept. to mid Oct.	Mid Aug. to mid Sept.
No reported experience	Medium to hard (canopy shaker) and medium (trunk shaker) harvestability; medium juicing; mostly as single berries
Medium spacing; bilateral cordon training with spur pruning	Medium spacing; trained to bilateral cordons and spur pruned. Spur selection may be difficult due to tendency of shoots to bend near base.
Low bunch rot potential and naturally of high acidity and relatively low pH; produces fruity wines of high color and tannin content, sometimes astringent	Berry shrivel from early maturity during hot weather prompts harvest. Subject to delayed growth in spring following years of poor wood maturity.

Minor Wine Grape Varieties
in California—*White*

	Arneis	Marsanne
Origin	Piedmont region of Italy where it is experiencing revival	Northern Rhône, France
Use	For delicate, distinctive, varietal table wines	For light table wines, usually blended, such as with Roussanne and Viognier
Clusters	Small to medium; broad conical, compact; medium-long peduncles	Medium; long cylindrical with broad shoulders or wings, well-filled to compact; medium-length peduncle
Berries	Small; round to short oval; yellow	Medium; round; yellow to amber
Leaves	Medium; 3-lobed with large superior lateral and apical lobes; U-shaped petiolar sinus; medium-long, sharp teeth; light, short hair, some tufted on lower leaf surface	Medium; 3- to 5-lobed with reduced inferior petiolar sinus or overlapping lobes, tightly closed lyre-shaped petiolar sinus due to overlapping lobes; inferior lateral lobes smaller and shoulder or wing-like; short, broad teeth; leaf surface rough, bullate and puckered; sparse tufted hair on lower leaf surface
Shoot tips	Wooly with rose margins; yellow-green with light-rose highlights	Felty white with rose margin; young leaves green-yellow with bronze-red highlights
Growth	Trailing with long, thin shoots. Medium to high vigor; canopy can be dense due to lateral shoot development.	Spreading; high vigor with moderately dense canopy
Adaptability	Cool to warm regions	Cool to warm regions; unfavorable experiences in hot regions
Yield	4 to 8 t/a	6 to 11 t/a
Harvest period	Early Sept. to mid Oct.	Mid Sept. to early Oct.
Harvest method	No reported experience	Easy to medium, mostly as single berries; medium juicing; heavy canopy can interfere
Culture	Medium spacing; bilateral cordon training with spur pruning	Medium to wide spacing; bilateral cordon training with spur pruning
Other characteristics	Compact clusters are susceptible to bunch rot. Easily overcropped due to good productivity with moderate vigor. Wide genetic variability in Italian vineyards; smaller and less compact cluster clones preferred.	Susceptible to powdery mildew; bunch rot and berry cracking are problematic with compact clusters and heavy canopies; of low fruit acidity and lacks varietal character in warm to hot climates

Orange Muscat	Sylvaner
Italy, known as Moscato Fior d'Arancio	Probably Austria; a leading variety of Central Europe, especially in the Moselle and Alsace regions
For premium dessert muscat wines	For delicate, distinct varietal table wines
Medium; conical with broad shoulders and cylindrical core, compact; medium-length peduncles	Small; cylindrical to short conical, well-filled to compact; short peduncles
Medium; round; yellow slight ambering when ripe; mild muscat flavor	Medium; round; green-yellow with prominent brown stylar scar and lenticels
Medium; almost entire to 3-lobed; narrow to closed U-shaped petiolar sinus; long, wide teeth; glabrous on the lower leaf surface	Small to medium; entire to slightly 3-lobed; U-shaped petiolar sinus (V-shaped at base); rounded teeth; glabrous lower leaf surface
Cobwebby; young leaves green with bronze-red patch, glabrous	Downy and white; young leaves yellow-green
Semi erect; medium vigor with moderately open canopy	Semi-upright; medium vigor; lateral shoot growth may result in dense canopies by midsummer
Warm to hot regions; poor tolerance to salts and boron when grown on own roots	Cool regions
7 to 12 t/a	4 to 6 t/a
Mid Aug. to mid. Sept.	Early to late Sept. Prompt harvest avoids low acid musts.
No reported experience	Easy to medium harvestability, mostly as single berries; light juicing
Close to medium spacing; bilateral cordon training with spur pruning; easily attained high cluster numbers	Medium spacing; bilateral cordon training with spur pruning; cane pruning tends to overcrop and weaken vines
Fruit readily attains high sugar levels above 24° Brix; exposed clusters tend to amber and raisin; strong, pleasant muscat flavor	Susceptible to bunch rot on deep, fertile soils and with fall rains; early ripening usually precludes this. Susceptible to powdery mildew. Spring frost recovery is fair to good.

Minor Wine Grape Varieties
in California—*White* (continued)

	Symphony	**Trousseau gris** (incorrectly Grey Riesling in California)
Origin	Cross of Grenache gris × Muscat of Alexandria by H.P. Olmo; UC patent 5013 in 1983	Northeastern France, where it is used in blends
Use	For fruity, dry to sweet muscat-flavored wines, including sparkling	Mostly for well-balanced but not distinctive varietal table wines in California
Clusters	Medium to large; broad conical, well-filled to compact; medium-length peduncles; very fruitful	Small; cylindrical, well-filled to compact; short to medium peduncles
Berries	Medium; round; yellow at maturity; muscat flavor	Small; oval; light purple-black, tan-red in sun
Leaves	Medium; 3-lobed, with mostly closed U-shaped petiolar sinus; reduced inferior lateral sinuses; short, sharp teeth; moderately dense hair on lower leaf surface	Medium; 3-to 5-lobed with shallow inferior lateral sinuses, often appear more or less entire and circular; narrow to closed U-shaped petiolar sinus; bullate surface and puckered; rounded teeth; light tufted hair on lower leaf surface.
Shoot tips	Wooly and white with slight pink margin; young leaves green and downy	Felty; young leaves yellow and downy
Growth	Semi-upright; vigorous and with dense canopy	Upright; very vigorous with thick canes and rather short internodes; tendency for dense canopy
Adaptability	Cool to hot regions	Cool to warm regions
Yield	7 to 10 t/a	5 to 8 t/a
Harvest period	Early to late Sept.	Late Aug. to early Sept.
Harvest method	Moderately easy with trunk shaker	Easy harvestability, removed as single berries; light juicing. Dense foliage may interfere.
Culture	Medium to wide spacing. Bilateral or quadrilateral cordon-trained with short, 2-node spurs; quadrilateral cordons reduce canopy congestion but may require some cluster thinning due to high productivity.	Medium spacing; train to bilateral or quadrilateral cordons with relatively high spur numbers due to medium-small clusters and high vigor. Leaf removal is often necessary.
Other characteristics	Tight clusters are prone to bunch rot. Good recovery from spring frosts.	Susceptible to Botrytis bunch rot and powdery mildew, especially with dense canopies. Early ripening contributes to bird feeding damage. Overcropping delays harvest, resulting in poor sugar to acid balance.

Ugni blanc
(St. Emillion, Trebbiano)

Probably Italy, although most widely grown in France

Most important brandy (Cognac) variety in France; also for blending of table wine

Medium to large; long conical, well-filled to compact; medium-length peduncle; often with a fork at base of rachis

Medium; round to oblate; yellow; large, scattered lenticels common

Medium to large; mostly 3-lobed with smaller and overlapping inferior lateral sinuses; overlapped U-shaped petiolar sinus; long, sharp teeth; dense tufted hair on lower leaf surface

Felty; young leaves yellow

Upright; high vigor and medium open canopy

Warm regions

10 to 13 t/a

Early to late Sept.

Medium harvestability and heavy juicing with canopy shaker

Medium to wide spacing; bilateral cordon training with spur pruning. Considered easy to grow, with low bunch rot susceptibility and ability to remain on vine without breakdown; easy to hand harvest.

Somewhat tolerant of powdery mildew and Botrytis rot but susceptible to Eutypa dieback. Shoots easily broken off in wind. Low acid, neutral wine.

Glossary

Ampelography. The art of identifying grape varieties by observation of leaf and fruit characteristics; also a compilation of morphological descriptions of grape varieties.

Anther. The pollen-bearing part of a stamen.

Anthesis. The time of full bloom in a flower, just after the calyptra (the united petals) falls.

Anthocyanin. A plant pigment that gives leaves and fruit a red, blue, or purple color.

Arms. The portion of the cordon or trunk from which the fruiting units arise.

Basal bud. *See* buds.

Berry set. The period after bloom (anthesis) when most of the flowers have dropped and the pistils of the pollinated flowers begin to develop into berries.

Bilateral cordon. *See* vine training.

Blade. The expanded portion of a leaf.

Blind buds. Nodes on spurs or canes from which there is no budbreak in the spring.

Bloom. The waxy coating on a mature grape berry, which often gives a frosted appearance to dark colored varieties.

Botrytis bunch rot. Fungal disease of the clusters (and occasionally shoots and leaves) caused by *Botrytis cinerea* encouraged by cool moist weather. Often leads to sour rot and cluster collapse in California.

Buds.

    Basal buds. Dormant buds that develop in the axils of bracts at the base of a shoot. Internodal elongation is minimal so buds appear to be in a whorl around the base of the shoot or cane.

    Count bud. *See* Count node.

    Dormant bud. The rounded organ on a cane or maturing shoot containing microscopic shoots and clusters, within protective scales. It contains a larger, primary bud and two smaller secondary buds and normally develops during the spring following initiation. Also known as an "eye."

Latent bud. Usually a dormant bud, located on wood more than one year old.

Lateral bud. The pointed bud in each leaf axil that gives rise to lateral shoots or abscises during the year of production. The dormant bud is produced in the basal node of the lateral bud. Also known as the "prompt" bud.

Bullate. A bumpy, quilted leaf surface where the tissue puckers upwards between the minor veins.

Calyptra. The fused petals of the grape that fall off the flower at anthesis, often known as the "cap."

Cane. A mature, lignified (woody) shoot. Can also be a long fruiting unit retained at pruning for fruit production, generally 12 to 15 nodes long.

Cane pruning. Pruning style that leaves 12- to 15-node canes for fruit production. Used with some varieties that do not have fruitful basal buds or where it is required by climate or training style.

Canopy management. Training of the vine to provide the light environment and leaf surface area to bring the crop and vegetation into balance to assure optimum fruit maturity and quality.

Canopy shaker harvester. Mechanical grape harvester that uses paired banks of fiberglass rods that pivot horizontally to strike and shake the canopy. Fruit is removed by the horizontal shaking motion and some direct contact of the rods to the fruit zone. Some harvesters have adjustable heads that allow the rods to strike at an upward angle. The fiberglass rods may be straight or bowed. Canopy shakers are sometimes referred to as pivotal striking or horizontal-rod harvesters.

Cap. *See* Calyptra.

Cap fall. *See* Anthesis.

Cap stem. The pedicel of the flower and the stem attachment of the berry to the rachis.

Certified Planting Stock. Grapevine propagation material that has been tested for known virus diseases and verified to be true to variety under a certification and registration program regulated by the California Department of Food and Agriculture.

Chlorosis. Yellowing or blanching of green leaves due to nutrient deficiencies, disease, or other factors.

Clone. A group of vines of a uniform type propagated vegetatively and tracing back to an original, selected mother vine.

Compatibility. Ability of the scion and stock to unite in grafting and form a healthy and durable union.

Cordon. The horizontal branch of the vine usually attached to a support wire and from which the fruiting units form.

Coulure. (French) Abscission of flowers before or during bloom, resulting in poorly filled clusters.

Count node. Number of dormant nodes on a spur or cane, not including basal buds. First count node is separated by ¼ inch or more from the basal bud below. Transitional forms may make determining first count bud difficult. Sometimes called a "count bud" when referring to the dormant bud at each node.

Degrees Brix (° Brix). A measure of the total soluble solid content of grapes, approximately the percentage of grape sugars in the juice.

Dentate. Teeth on the leaf margin.

Entire leaves. Leaves without lateral sinuses or these sinuses overlap to appear closed.

ENTAV. Etablissement National Technique pour l'Amélioration de la Viticulture.

Eutypa dieback. Severe fungal disease of grapevines caused by *Eutypa lata*. Spores of this fungus enter pruning wounds during rainy winter weather, and it slowly kills spurs, arms, and cordons.

Eye. *See* Bud, dormant.

Fasciation. The flattening of the stem and splitting of a shoot, usually associated with multiple buds in one area as a result of dramatically shortened internodes. Can be symptomatic of fanleaf degeneration.

FPS. Foundation Plant Service at the University of California, Davis (formerly Foundation Plant Materials Service); offering certified planting stock, disease testing, importation, and variety identification services for horticultural crops.

FPS selection. Identified at FPS both by variety name and selection number.

Fruit set. The stage of cluster development following the drop of unfertilized berries about two weeks after bloom. The small, retained berries are said to be "set." The term "shatter" is often used to describe the drop of unfertilized berries.

GDC (Geneva Double Curtain) systems. *See* Trellising and Canopy Management section.

Glabrous. Smooth leaves, not rough, pubescent, or hairy.

Goblet. The form of the arms in head training with spur pruning, resembling a vase or goblet.

Harvestability. Comparative ratings of "easy," "medium," and "hard" with machine harvesting are used in this publication. These describe the ease of fruit removal with force, or the force required for fruit removal. The ratings were derived from interviews with and survey results from wine grape industry representatives who had machine harvesting experience.

Head. The portion of a trunk where arms or cordons originate.

Head training (vertical cordon). *See* Vine training.

Herbaceous. Green, soft tissue, not woody.

INRA. Institut National de la Recherche Agronomique.

Internode. The portion of a shoot or cane between two adjacent nodes.

Juicing. Comparative rating of "light," "medium," and "heavy" with machine harvesting are used in this publication. These describe the relative amount of free juice that appears on the leaves and other vine parts and in the harvested fruit. Minimal juicing is always preferred.

Lateral. A branch of the main axis of the shoot or cluster.

Leaf scar. The scar left on the stem after a leaf falls.

Lenticel. A tiny, pore-like opening surrounded by corky tissue, often seen on grape berries and pedicels.

Lignified. The presence of woody or bark-like tissue on shoots (canes) or peduncles.

Lobe. Grape leaves are divided into five lobes centered around the five main leaf veins and separated by sinuses of varying degrees.

Lyre system. *See* Trellising and Canopy Management section.

Maturity. The later stage of grape berry ripeness, measured by the content of soluble solids (mostly sugars). Can also be evaluated by acids, color, and tannin structure.

Millerandage. (French) Fruit set that results in berries that are not uniform size, usually including shot berries.

MOG. Material Other than Grapes, such as leaves, stems, and other foreign materials in harvested grape loads.

Must. Unfermented grape juice, which contains seeds, stems, and skins.

Naked veins. When no tissue borders the first branch of the main veins along the petiolar sinus.

Node. The enlarged portion of the shoot or cane on which leaves, clusters, tendrils, and/or buds are located.

Own-rooted. A vine grown from a cutting that develops on its own root system; an ungrafted grapevine.

Pedicel. The small stem that attaches the berry to the rachis (cluster framework).

Peduncle. The stem that attaches the rachis (cluster framework) to the shoot/cane at the node.

Petiolar sinus. A cleft in the leaf margin at the attachment of the petiole.

Petiole. The stem that attaches the leaf blade to the shoot.

Pistil. The female part of the flower, consisting of a stigma, a style, and an ovary that becomes the berry.

Point mutations. When the genetic material of a single cell in a bud mutates in a manner that results in a change in the visual appearance of a cluster or shoot.

Powdery mildew. Fungal disease caused by *Uncinula necator* that forms a gray mold on all green parts of the vine. Must be controlled with systemic fungicides or sulfur.

Premium variety. A grape variety that produces high-quality wine of distinctive character.

Pruning.
    Cane. A pruning method in which canes (generally 12 to 15 nodes long) are retained as fruiting units.
    Spur. A pruning method in which short (generally 1 to 3 nodes long) spurs are retained as fruiting units.

Pubescent. Leaves and stems covered with short hairs.

Rachis. The branched cluster framework attached to the shoot or cane by the peduncle.

Rootstock. The understock upon which fruiting varieties are grafted. Rootstocks provide resistance to pests and diseases and tolerance to abiotic soil problems.

Rugose. A wrinkled or puckered leaf surface.

Scion. The fruiting variety grafted or budded onto a rootstock.

Second crop. Clusters initiated and produced in the current season.

Serrations. Teeth-like indentations at the margin of a leaf.

Set. Fruit set; the beginning stage of berry growth.

Shelling. Abscission of flowers before or in bloom. Occurs previous to the fruit set stage; is sometimes associated with excess vine vigor, excess nitrogen, or weather extremes that affect flower formation.

Shoot. The current season's stem growth that bears leaves, buds, clusters, and tendrils.

Shoot thinning. The removal of shoots from latent buds on cordons or arms or unwanted shoots from count nodes on spurs or canes. Also referred to as "shoot removal."

Shot berry. A very small berry that fails to develop to normal size, usually seedless and sometimes green.

Sinus. Gaps in the leaf margin that separate the leaf lobes. The lateral sinuses closest to the petiole are referred to as "inferior," and those near the apical lobe are "superior." Petiolar sinus is defined above.

Spur. A short fruiting unit of one-year growth usually consisting of one to three nodes and retained at pruning.

Still wine. Wine without the bubbles of sparkling wine.

Stock. *See* Rootstock.

Stylar scar. A small, corky area remaining on the apex of a berry after the style abscises following fertilization.

Sucker. Common viticultural usage describes a shoot arising from latent buds on the trunk or rootstock. Horticulturalists use the less common term "watersprout" for a shoot developing on the trunk.

Suckering. Common viticultural usage describes the removal of latent shoots (or "suckers") from trunk or rootstock.

Tendril. A slender, usually branched two or more times structure that coils around objects and supports the shoot. They originate opposite the leaves at the nodes and can differentiate into small clusters.

Terpenes. A class of organic compounds important as odorants of the highly aromatic muscat grapes and related varieties such as Riesling and Gewürztraminer.

Tomentum. Short to long hair often densely matted on the leaf or shoot surface.

Trunk. The main stem or body of a vine between the roots and the place where it divides to form branches.

Trunk shaker harvester. Mechanical grape harvester that shakes the trunk horizontally just below the vine cordon or head. Two rails apply energy to the trunk by oscillating sideways and with no direct contact to the canopy. The method relies heavily on efficient energy transfer through the vine aided by a stake at each vine. It results in more whole cluster removal than canopy shaker harvesters, but it is adaptable to fewer vineyard designs. Trunk shakers are sometimes referred to as pulsator harvesters.

Varietal character. Distinctive aroma or flavor characteristics of certain grape varieties that make them recognizable.

Variety. Botanically, a group of closely related plants within a species. A horticultural variety or race is correctly referred to as a cultivar, although variety is usually used in viticulture.

Veraison. The stage of development where berries begin to soften and/or color.

Vine training.

Bilateral cordon. The trunk is divided into two permanent horizontal branches each supported by a wire and extending in opposite directions. The fruiting positions originate off of these cordons.

Guyot. The trunk is topped at a fruiting wire, giving rise to one or several fruiting canes and renewal spurs. The canes may be tied down onto a horizontal wire or arched over several wires.

Head. A simple system of training in which the upright trunk is held up by a stake and short, permanent arms that bear spurs or canes branch off the trunk.

Quadrilateral cordons. The trunk is divided into two short branches that are attached to parallel wires running up and down the row. Each of these branches is divided into bilateral cordons.

Unilateral cordon. The trunk is trained to a single branch, which is supported on a horizontal wire and gives rise to the fruiting units.

VSP system. *See* Trellising and Canopy Management section.

Watersprout. *See* Sucker.

Wing. A well-developed basal cluster lateral that projects and is separated from the main body of the cluster.

Winkler Region. A way of classifying California's climatic grape-growing regions based on the heat summation of mean daily temperatures above 50°F and expressed as degree days. The heat summations of the climatic regions are
I: less than 2,500 DD (degree days)
II: 2,501 to 3,000 DD
III: 3,001 to 3,500 DD
IV: 3,501 to 4,000 DD
V: 4,001 or more DD

Bibliography

Alley, L., and D. Golino. 2000. Origins of the grape program at Foundation Plant Materials Service. American Journal of Enology and Viticulture, Proceedings of the 50th annual ASEV meeting, Seattle, Wa. AJEV 51: 222–230.

Amerine, M. A., and A. J. Winkler. 1944. Composition and quality of musts and wines of California grapes. Hilgardia 15: 493–673.

———. 1963. California wine grapes: Composition and quality of their musts and wines. California Agricultural Experiment Station Bulletin 794.

Anonymous. 1997. The Brooks and Olmo register of fruit and nut varieties. Alexandria, Va.: American Society of Horticultural Science Press.

———. 1997. Catalogue of selected wine grape varieties and clones cultivated in France. St. Pons de Thomieres, France: Ministry of Agriculture, Fisheries and Food, CTPS.

———. 2001. California grape register. Davis: Foundation Plant Materials Service, University of California. June.

———. 2002a. California grape acreage, 2001. Sacramento: California Agriculture Statistics Service, California Department of Food and Agriculture.

———. 2002b. Clonal aspects of wine-growing: Short course proceedings. Davis: University of California Extension.

Belfrage, N. 1999. Barolo to Valpolicella: The wines of Northern Italy. London: Faber and Faber.

Clarke, O., and M. Rand. 2001. Oz Clarke's encyclopedia of grapes. London: Harcourt.

Coombe, B. G., and P. R. Dry, eds. 1988. Viticulture, Volume I: Resources in Australia. Adelaide: Australian Industrial Publishers.

Flaherty, D. L., L. P. Christensen, W. T. Lanini, P. A. Phillips, and L. T. Wilson, eds. 1992. Grape pest management, 2nd ed. Oakland: University of California Division of Agriculture and Natural Resources Publication 3343.

Galet, P. 1998. Grape varieties and rootstock varieties. English Edition. France: Oenoplurimédia.

———. 2002. Grape varieties. London: Cassell Illustrated.

Golino, D. 2000. Trade in grapevine plant materials: local, national and worldwide perspectives. American Journal of Enology and Viticulture, Proceedings of the 50th annual ASEV meeting, Seattle, Wa. AJEV 51: 216–221.

Kasimatis, A. N., B. E. Bearden, and K. Bowers. 1979. Wine grape varieties in the North Coast counties of California. Berkeley: University of California Division of Agricultural Sciences Publication 4069.

Kasimatis, A. N., L. P. Christensen, D. A. Luvisi, and J. J. Kissler. 1980. Wine grape varieties in the San Joaquin Valley. Berkeley: University of California Division of Agricultural Sciences Publication 4009.

Kerridge, G., and A. Antcliff. 1996. Wine grape varieties of Australia. Collingwood, Victoria: CSIRO Publishing. Division of Horticulture.

May, P. 1994. Using grapevine rootstocks: The Australian perspective. Adelaide: Winetitles.

McKay, A., G. Crittenden, P. Dry, and J. Hardie. 1999. Italian wine grape varieties in Australia. Adelaide: Winetitles.

Mullins, M. G., A. Bouquet, and L.E. Williams. 1992. Biology of the grapevine. Cambridge: Cambridge University Press.

Muscatine, D., M. A. Amerine, and B. Thompson, eds. 1984. The book of California wine. Berkeley: University of California Press/Sotheby Publications.

Orffer, C. J., ed. 1979. Wine grape cultivars in South Africa. Cape Town: Human and Rousseau.

Pongrácz, D. P. 1983. Rootstocks for grapevines. Cape Town: David Phillip.

Rantz, J. M., ed. 1995. Proceedings of the international symposium on clonal selection. Davis: American Society for Enology and Viticulture.

Robinson, J. 1998. Vines, grapes and wines: The wine drinker's guide to grape varieties. London: Mitchell Beazley.

Smart, R., and M. Robinson. 1991. Sunlight into wine. Adelaide: Winetitles.

Sullivan, C. L. 1998. A companion to California wine. Berkeley: University of California Press.

Walker, M. A. 2000. UC Davis' role in improving California's grape planting materials. American Journal of Enology and Viticulture, Proceedings of the 50th annual ASEV meeting, Seattle, Wa. AJEV 51: 209–215.

Winkler, A. J., J. A. Cook, W. M. Kliewer, and L. A. Lider. 1974. General viticulture. Berkeley: University of California Press.

Wolpert, J. A., M. A. Walker, and E. Weber. 1992. Rootstock seminar proceedings: A worldwide perspective. Davis: American Society for Enology and Viticulture.

Index

cane pruning
    defined, 179, 181
    uses, 17, 19
canes
    defined, 179
    structure, 4
canopy management
    defined, 179
    See also pruning systems; training sys-
       tems; trellis systems
canopy shaker harvester, defined, 179
cap fall. *See* anthesis
caps. *See* calyptras
cap stems, defined, 179
Cardonnay variety, **44–49**
Carignane variety, 6, **40–43**, 131
Carnelian variety, 6, **169**
Carneros, ripening dates, 7
Centurion variety, 6, **169**
Certified Planting Stock
    defined, 180
    FPS selections, 8–10
Chardonnay variety, 6, **44–49**
Chateau Yquem Vineyard, 139
Chenin blanc variety, 6, **50–53**
chlorosis, defined, 180
Cinsaut variety, 6, **170**
clones
    defined, 8, 180
    diversity of, 9
    intellectual property issues, 9–10
    labeling and identification of, 8–11
    performance evaluation, 11
    selection criteria, 10–11
    See also specific clones and profiles of spe-
       cific grape varieties
cluster blight (Botrytis), of Syrah, 149
collar rot, of Rubired, 128
Colombard variety, 6, **54–57**
compatibility, defined, 180
cordons
    bilateral, defined, 179
    defined, 180
    quadrilateral, defined, 182
    unilateral, defined, 182
    See also training systems
Cot or Cote rouge varieties. *See* Malbec vari-
    ety
Couderc rootstock. *See* 1616C rootstock;
    3309C rootstock
coulure
    defined, 180
    of Malbec, 75, 77
count nodes, defined, 180
Crljenak Kasteljanski variety, 163
crown gall
    of Barbera, 27
    of Mission, 171

D
degrees Brix (° Brix), defined, 180
Delmas, Antoine, 87, 99
dentate leaves, defined, 180
Dijon clones, 109
diseases of plants
    black measles, 39, 101
    certified stock to control, 8–9, 11
    collar rot, 128
    crown gall, 27, 171
    downy mildew, 43
    esca, 93
    Phomopsis cane and leaf spot of, 57, 65,
       73, 101, 149
    See also bunch rot; Eutypa dieback;
       Pierce's disease; powdery mildew;
       virus diseases
Docetto variety, **170**
Dogridge rootstock, characteristics, 14–15
Dolcetto variety, 6, **170**

downy mildew, of Carignane, 43
Doyle, John, 25
drought tolerance of specific rootstocks, 12,
    14
Dureza variety, 147
Durif variety, 6, **58–61**

E
ENTAV (Etablissement National Technique
    pour l'Amélioration de la Viticulture)
    clonal diversity role, 9
    clone numbers, 10
    defined, 180
    See also profiles of specific grape varieties
ENTAV-INRA clones, 10
entire leaves, defined, 180
esca, of Mourvèdre, 93
Eutypa dieback
    of Cabernet Sauvignon, 39
    of Carignane, 43
    of Chenin blanc, 53
    of Colombard, 57
    defined, 180
    of Grenache, 73
    of Merlot, 89
    of Pinot Meunier, 117
    of Rubired, 128
    of Sauvignon blanc, 141
    of Tempranillo, 153
    of Ugni blanc, 177
    of Zinfandel, 166
eyes. *See* buds: dormant

F
fasciation, defined, 180
Foundation Plant Materials Service. *See* FPS
    (Foundation Plant Service)
Foundation Vineyard
    about, 8
    heritage field selections, 9
FPS (Foundation Plant Service)
    about, 8
    defined, 180
FPS selections
    about, 8–9
    defined, 180
Freedom rootstock
    for Barbera, 25
    for Burger, 29
    for Carignane, 41
    characteristics, 14–15
    for Chenin blanc, 51
    for Colombard, 55
    for Durif, 59
    for Grenache, 71
    for Malbec, 75
    for Malvasia bianca, 79
    for Merlot, 87
    for Muscat blanc, 95
    for Muscat of Alexandra, 100
    for Riesling, 119
    for Rubired, 127
    for Ruby Cabernet, 131
    for Sangiovese, 135
    for Syrah, 147
    for Tempranillo, 151
    for Viognier, 159
Freisa variety, 6, **171**
French Colombard variety. *See* Colombard
    variety
Fresno, ripening dates, 7
frost damage
    of Chardonnay, 49
    of Gewürztraminer, 69
    of Grenache, 73
fruit set, defined, 180, 181
Fumé or Fumé blanc varieties. *See* Sauvignon
    blanc variety

G
Galet, Pierre, 155
Gamay Beaujolais variety. *See* Gamay noir
    variety
Gamay noir variety, 6, **62–65**
Gamay variety. *See* Valdiguié variety
Garnacha variety. *See* Grenache variety
Geisenheim Research Institute, 9, 119–120
genetic mutations
    clones and, 8
    point, defined, 181
    point, of Pinot family, 107
Gewürztraminer variety, 6, **66–69**
giberellin sprays, 53
glabrous leaves, defined, 180
goblet training, defined, 180
Goheen, Austin, 8
Gouais blanc variety, 45, 63
graft incompatibilities
    of Cabernet Sauvignon, 38
    of Chardonnay, 46, 48
grape berries
    bloom, defined, 179
    cracking, of Marsanne, 174
    hens and chickens, 161
    maturity, defined, 181
    set, defined, 179
    size and shape, 5
    See also profiles of specific grape varieties
grape berries, shot
    defined, 181
    of Gamay noir, 65
grape berries, shrivel
    of Dolcetto, 170
    of Durif, 59, 61
    of Syrah, 149
    of Tinta Madeira, 173
grape clusters, 4
    attachment to cane, 5
    size and shape, 5
    See also profiles of specific grape varieties
grape growers, history, 3
grape varieties
    defined, 182
    labeling and identification of, 8–11
    premium, defined, 181
    selection criteria, 8–9
    See also clones; *profiles of specific grape*
       varieties
grapevines, structure, 4, 5
Grenache noir variety. *See* Grenache variety
Grenache variety, 6, **70–73**
guyot training, defined, 182

H
Harazthy, Agostin, 99
Harmony rootstock
    for Barbera, 25
    for Carignane, 41
    characteristics, 14–15
    for Chenin blanc, 51
    for Colombard, 55
    for Grenache, 71
    for Muscat blanc, 95
    for Muscat of Alexandra, 100
    for Rubired, 127
    for Syrah, 147
harvestability, defined, 180
harvesting. *See also profiles of specific grape*
    varieties
    canopy shaker harvester, 179
    mechanization, 17, 19, 21
head of vine, defined, 180
Healdsburg, ripening dates, 7
hedging, 18
hens and chickens, 161
herbaceous, defined, 180
Hewitt, William, 8
Hilgard, Eugene, 3